METAL HALIDE PEROVSKITES

Synthesis, Properties and Applications

METAL HALIDE PEROVSKITES

Synthesis, Properties and Applications

Jin Zhong Zhang

University of California Santa Cruz, USA

Zhiguo Xia

South China University of Technology, China

Qi Pang

Guangxi University, China

World Scientific

NEW JERSEY · LONDON · SINGAPORE · BEIJING · SHANGHAI · HONG KONG · TAIPEI · CHENNAI · TOKYO

Published by

World Scientific Publishing Co. Pte. Ltd.

5 Toh Tuck Link, Singapore 596224

USA office: 27 Warren Street, Suite 401-402, Hackensack, NJ 07601

UK office: 57 Shelton Street, Covent Garden, London WC2H 9HE

Library of Congress Control Number: 2022945018

British Library Cataloguing-in-Publication Data
A catalogue record for this book is available from the British Library.

METAL HALIDE PEROVSKITES
Synthesis, Properties and Applications

ISBN 978-981-125-741-4 (hardcover)
ISBN 978-981-125-742-1 (ebook for institutions)
ISBN 978-981-125-743-8 (ebook for individuals)

For any available supplementary material, please visit
https://www.worldscientific.com/worldscibooks/10.1142/12877#t=suppl

Desk Editor: Nur Syarfeena Binte Mohd Fauzi

Typeset by Stallion Press
Email: enquiries@stallionpress.com

We wish to dedicate this book to our families, teachers, and mentors.

Preface

Metal halide perovskites (MHPs), including both organo-metal halide perovskites (OMHPs) and all inorganic metal halide perovskites (IMHPs), have attracted tremendous attention in the last decade, partly due to the outstanding performance of OMHPs such as methyl ammonium lead iodide ($CH_3NH_3PbI_3$ or $MAPbI_3$) that exhibits outstanding photovoltaic (PV) performance, exceeding 20% as certified by the National Renewable Energy Lab (NREL). The strong PV performance is due to some unique electronic and optical properties of the $MAPbI_3$. Subsequently, MHPs have been investigated for applications in areas including light emitting diodes (LEDs), sensing, and imaging.

While there are many thousands of papers on MHPs and a few edited books covering various aspects of MHPs including different applications, to date, there has been no book with a focus on systematically covering synthesis, fundamental properties as well as applications of MHPs in a single monograph. This book is intended to fill this gap and to provide beginning researchers with the necessary background information about MHPs and practitioners a good resource for useful information including highlights of some of the important recent developments.

The book covers the synthesis, properties, and applications of MHPs and each subject is presented in terms of bulk and nanostructures to make it easier to cover and follow. There have been some excellent recent reviews and edited books related to the subjects, and many are referenced in this book.

Since research on MHPs is still growing very quickly, it is not possible to cover all aspects of MHPs or all the recent or past developments. The objective here is to introduce MHPs with an emphasis on their fundamental properties and emerging applications. Examples used in the book are completely the author's discretion. However, whenever possible, the authors made efforts to use recent and significant examples that are well established, to the best judgment of the author.

We welcome feedback from readers and will attempt to incorporate them in future editions of this book if there is such an opportunity.

Jin Zhong Zhang
Santa Cruz, CA, USA
zhang@ucsc.edu

Zhiguo Xia
Guangzhou, Guangdong, China
xiazg@scut.edu.cn

Qi Pang
Nanning, Guangxi, China
pqigx@163.com
July 2022

Acknowledgment

We would like to thank our mentors and many colleagues, collaborators, postdoctors, and students who have helped directly or indirectly with the writing of this book, through discussion, collaboration, and research work. An incomplete list of people to whom we wish to express our gratitude includes Sara Bonabi, Guozhong Cao, Daimei Chen, Yibo Chen, Jason Cooper, Hai-Lung Dai, Elder de la Rosa, Mostafa El-Sayed, Xiaosheng Fang, Ghada Farouk, Melissa Guarino-Hotz, Jinghua Guo, Ting Guo, Kai Han, Charles B. Harris, Greg Hartland, Eric J. Heller, Yung Jung Hsu, Jianhua Hu, Yun Hang Hu, Dan Imre, Bo Jiang, Jince Jin, Prashant Kamat, Dongling Ma, Shuit-Tong Lee, Stephen Leone, Can Li, Dehui Li, Jinghong Li, Xueming Li, Yan Li, Yat Li, Sarah Lindley, Gang-yu Liu, Jun Liu, Li Liu, Tzarara Lopez-Luke, Liqiang Lu, Binbin Luo, Xiaojun Lv, Haixia Ma, Yuan Ping, Ying-Chih Pu, Oleg Prezhdo, George Schatz, Greg Scholes, Ben Schwartz, Adam Schwartzberg, Jianying Shi, Evan Vickers, Jianwu Wei, Changchun Wang, Hongmei Wang, Lei Wang, Zhenyang Wang, Ke Xu, Dongfeng Xue, Kui Yu, Ping Yu, Peng Zhang, Zhongping Zhang, Ruosheng Zeng, Haizheng Zhong, Bingsuo Zou, and Shengli Zou.

We are grateful to our families and friends for their support, love, and understanding.

We wish to thank the book editor, Nur Syarfeena Binte Mohd Fauzi, for her wonderful and professional assistance.

Contents

Chapter 1

Introduction

Metal halide perovskites (MHPs), including both organic–inorganic hybrid and all-inorganic, have attracted significant attention in recent years because of their intriguing properties, unique functionalities, as well as promising applications in emerging technologies including photovoltaics (PV), light-emitting diodes (LEDs), detectors, sensors, and displays. Most notable is their outstanding PV performance in solar energy conversion, with power conversion efficiency (PCE) for PV pushed from 9.7% in 2014 to 25.2% in 2020 [1–4]. MHPs based on lead and iodide have a number of *advantages and unique features*. These include (1) strong absorption in the UV-visible region (<750 nm), ideal for solar energy conversion [5,6], (2) high charge-carrier mobility and long diffusion length (>1 μm) [7,8], (3) layered structures allowing for easy control of the bandgap by substituting metal or iodide with other elements or by changing the organic cations, (4) easy processing in solution or solid environment, and (5) potentially low cost for large-scale applications. One downside is that Pb is toxic; however, this can be addressed by proper packaging and handling of the products at the end of their lifetime [6–9]. Research is also being undertaken to find a replacement for Pb with less toxic metals, such as Bi and Sn [10].

Figure 1.1 shows the growth in the citation of the first paper on perovskite solar cells published by Kojima and co-workers [11]. This indicates the significant growth in research activity in this field.

Since the first publication in 2009, major achievements have been made. Figure 1.2 shows some major milestones and related timelines.

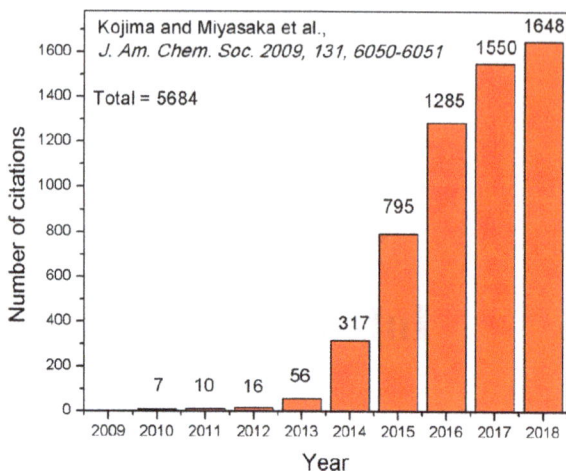

Figure 1.1. Year-wise citation history of the first paper on perovskite solar cells by Kojima *et al.* (2009), based on data obtained from Clarivate Analytics. Reproduced from Ref. [11].

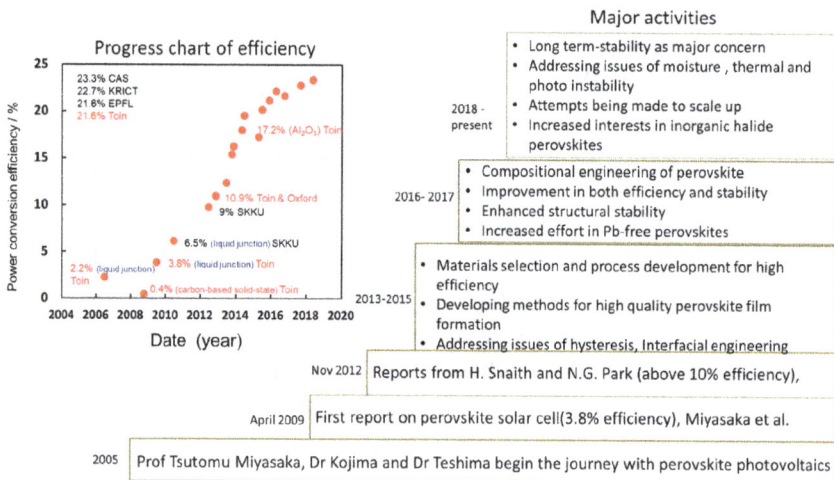

Figure 1.2. Right: Major milestones and activities of progress in perovskite solar cells from 2005 to 2018. Left: Power conversion efficiency progress chart. Reproduced from Ref. [11].

To date, the published and NREL-certified PCE has been over 25%, which is truly outstanding [4].

Besides PV, MHPs have shown good potential for LED applications [12–15]. The highest external quantum efficiency (EQE) has been reported in 2018 for PLEDs fabricated using perovskite quantum dots (PQDs) by anion exchange from pristine $CsPbBr_3$ using halide-anion-containing alkyl ammonium and aryl ammonium salts [16].

MHPs for sensing and detection of radiation such as light and X-rays. Photodetectors or radiation detectors are one of the latest additions in an expanding list of applications of MHPs. One key aspect is that MHPs can accommodate heavy elements while being able to form large, high-quality crystals and polycrystalline films, making them promising for X-ray and γ-ray detector technologies [17]. Figure 1.3 shows some examples of metal halide perovskites commonly used in radiation detectors, the linear attenuation coefficient of $CsPbI_3$, $MAPbBr_3$, CdTe, Se, and TlBr versus photons energy, and the absorption coefficient and length versus photon energy from visible to hard X-rays for $MAPbI_3$ [17].

MHPs are also promising for spintronics and spin-optotronics applications based on their spin properties including spin-orbital coupling (SOC) that is particularly prominent in heavy elements, such as Pb and I. For instance, a recent study has demonstrated spin-polarized carrier injection into methylammonium lead bromide ($CH_3NH_3PbBr_3$ or $MAPbBr_3$) films from metallic ferromagnetic electrodes in spintronic-based devices [18]. Figure 1.4 show the device structure, giant magnetoresistance (GMR) and Hanle effect in a $CH_3NH_3PbBr_3$ ($MAPbBr_3$)-based spin valve (SV) [18]. In addition, a 'spin light-emitting diode' results in circularly polarized electroluminescence emission. This study demonstrates the promise of hybrid perovskites for spin-related optoelectronic applications.

While bulk MHPs have shown great promise for applications, nanostructured MHPs, such as quantum dots (QDs) or nanocrystals (NCs), have also attracted considerable attention lately. Perovskite QDs (PQDs) or perovskite NCs (PNCs) afford additional advantages, including tunable electronic and optical properties by controlling size and shape, critical for applications such as display and imaging [13,14]. They are also potentially easier to process in solution using low-cost techniques such as inkjet printing

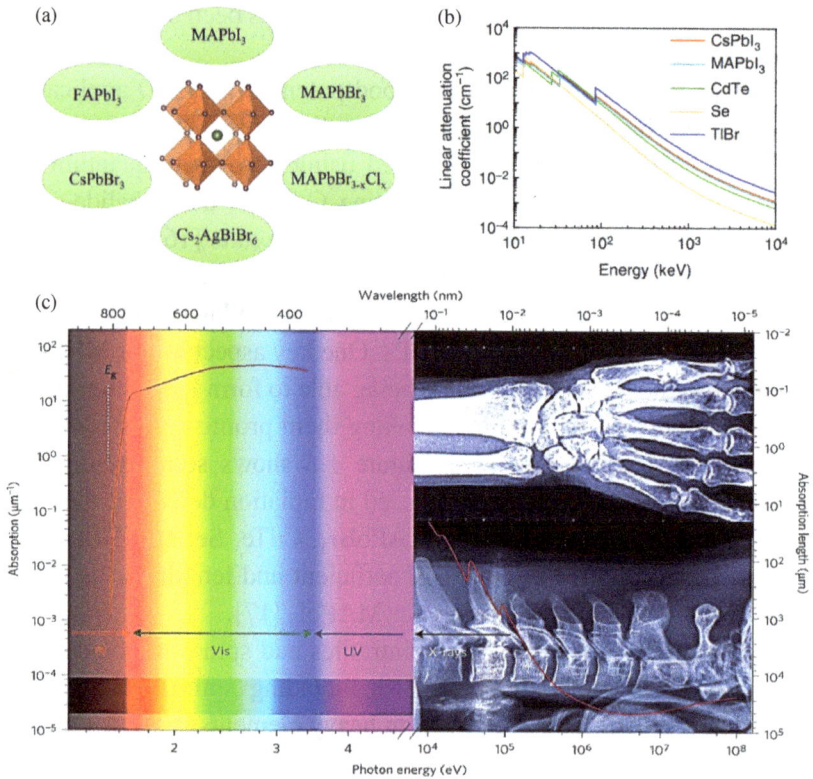

Figure 1.3.   (a) Metal halide perovskites commonly used in radiation detectors. (b) The linear attenuation coefficient of $CsPbI_3$, $MAPbBr_3$, CdTe, Se, and TlBr versus photons energy. (c) Absorption coefficient and length versus photon energy from visible to hard X-rays for $MAPbI_3$. Reproduced from Ref. [17].

and are more compatible with flexible substrates [19]. Furthermore, they serve as a good model system for studying the effect of surfaces due to their large surface-to-volume (S/V) ratio. One limitation with PQDs is their typically lower charge transport due to limited electronic coupling between PQDs in the solid forms. If the charge transport can be improved for PQDs, major advances can be expected for device applications [20]. Figure 1.5 shows an illustration of the synthesis process for $CH_3NH_3PbBr_3$ PQDs and their different colors by varying the halide along with their PL and EL spectrum as well as related CIE coordinates [21]. Both particle size and chemical

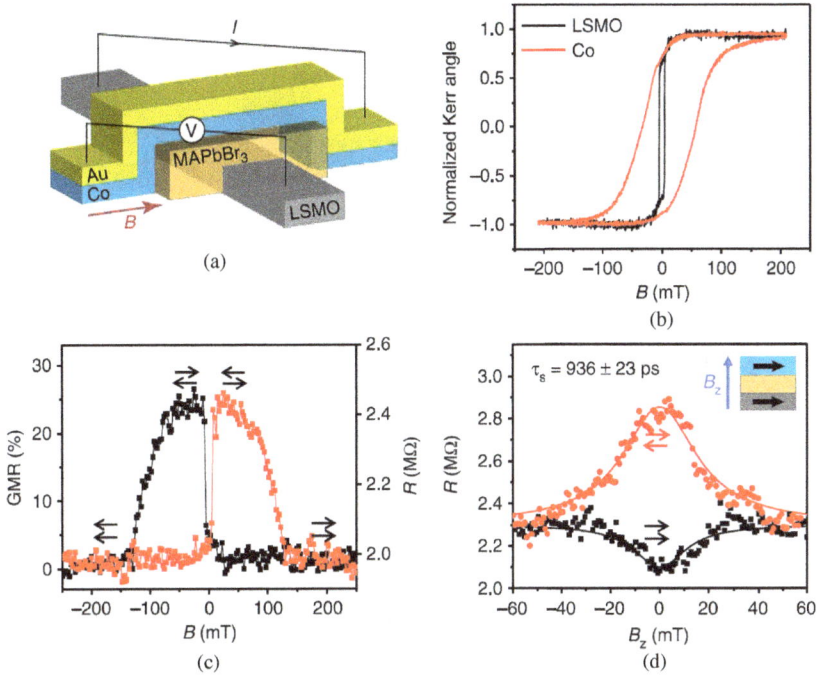

Figure 1.4. GMR and Hanle effect in MAPbBr$_3$-based spin valve. (a) Schematic diagram of a LSMO (La$_{0.63}$Sr$_{0.37}$MnO$_3$)/MAPbBr$_3$/Co SV device structure. (b) MOKE (magneto-optics Kerr effect) response of the LSMO and Co ferromagnetic electrodes as measured *in situ* in the SV device at 10 K. (c) GMR response of the SV measured at 10 K and applied bias voltage, $V = 0.1$ V. The obtained maximum GMR value (GMR$_{max}$) is 25%. The red and black lines represent magnetic field sweep up and down, respectively. The arrows show the mutual magnetization direction of the two FM electrodes. (d) Hanle effect of the GMR measured when a magnetic field, B$_z$ perpendicular to the FM electrodes is applied at both parallel and antiparallel magnetization configurations. The solid lines are fits, from which a spin lifetime $\tau_s = 936 \pm 23$ ps is extracted. Reproduced from Ref. [18].

composition can be used to tune the optical properties of the PQDs in a wide spectral range, which is highly desired for many applications.

Recently, there has been growing interest in 2D MHPs partly because some of them have demonstrated higher stability than 3D MHPs [22,23]. There is also increasing attention to metal halide double perovskites (MHDPs) that exhibit some unique properties compared to the normal single crystal MHPs [24,25]. Another major direction of increasing

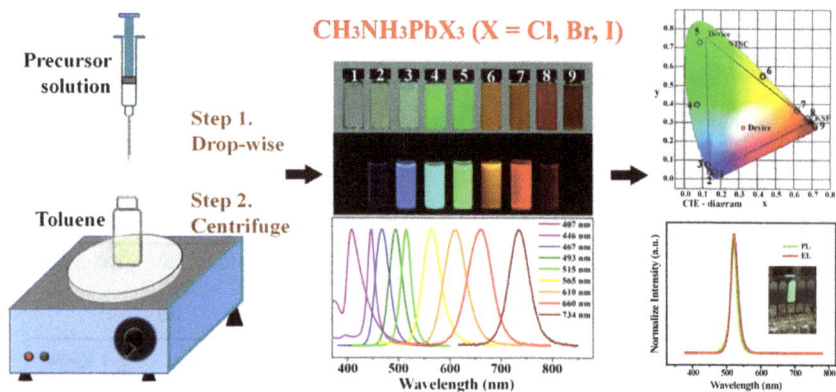

Figure 1.5. $CH_3NH_3PbBr_3$ PQDs: Illustration of synthesis methods (left), photograph and photoluminescence PL spectra (middle), CIE color coordinates (upper right), and normalized EL and PL spectra (lower right). Reproduced from Ref. [21].

attention is lead-free MHPs in which Pb is replaced by less toxic metals to reduce the overall toxicity of the MHPs that is a critical factor to consider for applications [26,27].

Parallel to experimental research, substantial theoretical and computational research has been conducted on MHPs in the last few years, most based on quantum mechanical calculations such as density functional theory (DFT) or different variations. These studies can validate experimental results, predict new structures or properties, and help explain experimental observations. Thanks to advances in computers and computational methods, highly sophisticated calculations of relatively large systems can nowadays be carried out, with good accuracy, which can be compared well with experimental results. One of the most important properties of interest is the electronic structure of the MHPs including energy levels of defects and surface states that strongly affect the properties and functionalities of the materials. For example, *ab initio* molecular dynamics simulations have been used to understand the charge hopping mechanism in zero-dimensional (0D) Cs_4PbBr_6 perovskite [28]. Figure 1.6 shows a schematic diagram of charge carrier hopping paths and dimer model for calculating the charge carrier mobility of 0D Cs_4PbBr_6 along with charge density distributions and charge density mapping of conduction band

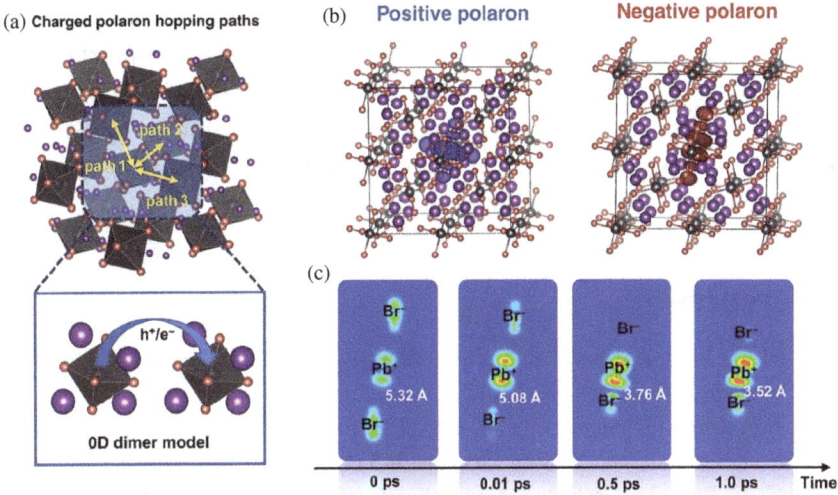

Figure 1.6. Illustration and calculation polaron transport and localization. (a) Schematic diagram of charge carrier hopping paths and dimer model for calculating the charge carrier mobility of 0D Cs_4PbBr_6. (b) Charge density distributions for a $2 \times 2 \times 2$ Cs_4PbBr_6 super-cell with shortened Pb–Br distances (positive polaron) and enlarged Pb–Br distances (negative polaron) within the central $[PbBr_6]^{4-}$ octahedron. (c) Charge density mapping of CBM of the central octahedron at selected times (0, 0.01, 0.05, 0.1, 0.5, and 1.0 ps) upon release from the initial state. The initial state at 0 ps was set as the negative-polaron state and had the longest Pb–Br bonds in the central octahedron. The selected Pb–Br bond lengths are indicated in each time delay. Reproduced from Ref. [28].

minimum (CBM) of the central octahedron at selected times (0, 0.01, 0.05, 0.1, 0.5, and 1.0 ps).

The above examples illustrate the excitement and promise of MHPs in both fundamental research and emerging technologies. However, there are still many challenges facing the field, especially in terms of long-term stability of the material and devices. Therefore, further research and development are still needed. The field is expected to continue to grow at least for another decade and technological applications will ultimately become mature and blossom.

This book consists of three major parts: synthesis (Chapter 2), properties (Chapter 3) and applications (Chapter 4). In each chapter, we cover the fundamental concepts and important recent examples, with the goals of giving newcomers a chance to gain familiarity with the basics of the field and of updating practitioners with the latest developments. Of course, given the limited space in the book and the vast number of publications and fast development of the field, it is not possible to be comprehensive. Rather, we have to be quite selective, and the selections are illustrative and reflect the authors' discretions.

References

[1] M.M. Lee, J. Teuscher, T. Miyasaka, T.N. Murakami, and H.J. Snaith, *Science* 338 (6107), 643 (2012).

[2] H.P. Zhou, Q. Chen, G. Li, S. Luo, T.B. Song, H.S. Duan, Z.R. Hong, J.B. You, Y.S. Liu, and Y. Yang, *Science* 345 (6196), 542 (2014).

[3] N.-G. Park, M. Grätzel, T. Miyasaka, K. Zhu, and K. Emery, *Nat. Energy* 1, 16152 (2016).

[4] E.J. Juarez-Perez and M. Haro, *Science* 368 (6497), 1309 (2020).

[5] H.S. Kim, S.H. Im, and N.G. Park, *J. Phys. Chem. C* 118 (11), 5615 (2014).

[6] N.G. Park, *J. Phys. Chem. Lett.* 4 (15), 2423 (2013).

[7] S.D. Stranks, G.E. Eperon, G. Grancini, C. Menelaou, M.J.P. Alcocer, T. Leijtens, L.M. Herz, A. Petrozza, and H.J. Snaith, *Science* 342 (6156), 341 (2013).

[8] G.C. Xing, N. Mathews, S.Y. Sun, S.S. Lim, Y.M. Lam, M. Gratzel, S. Mhaisalkar, and T.C. Sum, *Science* 342 (6156), 344 (2013).

[9] G. Niu, W. Li, F. Meng, L. Wang, H. Dong, and Y. Qiu, *J. Mater. Chem. A* 2 (3), 705 (2014).

[10] T. Miyasaka, *Chem. Lett.* 44 (6), 720 (2015).

[11] A.K. Jena, A. Kulkarni, and T. Miyasaka, *Chem. Rev.* 119 (5), 3036 (2019).

[12] Y.H. Kim, H. Cho, J.H. Heo, T.S. Kim, N. Myoung, C.L. Lee, S.H. Im, and T.W. Lee, *Adv. Mater.* 27 (7), 1248 (2015).

[13] J.W. Wei, F.R. Huang, S.N. Wang, L.Y. Zhou, Y.L. Xin, P. Jin, Z. Cai, Z.D. Yin, Q. Pang, and J.Z. Zhang, *Mater. Res. Bull.* 106, 35 (2018).

[14] M. Zhou, M.T.S. Mok, H. Sun, A.W. Chan, Y. Huang, A.S.L. Cheng, and G. Xu, *Oncogene* 36 (29), 4135 (2017).

[15] P.P. Du, L. Gao, and J. Tang, *Front. Optoelectron.* 13 (3), 235 (2020).

[16] T. Chiba, Y. Hayashi, H. Ebe, K. Hoshi, J. Sato, S. Sato, Y.J. Pu, S. Ohisa, and J. Kido, *Nat. Photon.* 12 (11), 681 (2018).

[17] G. Kakavelakis, M. Gedda, A. Panagiotopoulos, E. Kymakis, T.D. Anthopoulos, and K. Petridis, *Adv. Sci.* 7 (22), 33 (2020).

[18] J.Y. Wang, C. Zhang, H.L. Liu, R. McLaughlin, Y.X. Zhai, S.R. Vardeny, X.J. Liu, S. McGill, D. Semenov, H.W. Guo, R. Tsuchikawa, V.V. Deshpande, D.L. Sun, and Z.V. Vardeny, *Nat. Commun.* 10, 6 (2019).

[19] D.M. Balazs, N. Rizkia, H.H. Fang, D.N. Dirin, J. Momand, B.J. Kooi, M.V. Kovalenko, and M.A. Loi, *ACS Appl. Mater. Interfaces* 10 (6), 5626 (2018).

[20] A. Swarnkar, A.R. Marshall, E.M. Sanehira, B.D. Chernomordik, D.T. Moore, J.A. Christians, T. Chakrabarti, and J.M. Luther, *Science* 354 (6308), 92 (2016).

[21] G.L. Yang and H.Z. Zhong, *Chin. Chem. Lett.* 27 (8), 1124 (2016).

[22] T. Zhu, Y. Yang, K. Gu, C. Liu, J. Zheng, and X. Gong, *ACS Appl. Mater. Interfaces* 12 (46), 51744 (2020).

[23] K. Leng, L. Wang, Y. Shao, I. Abdelwahab, G. Grinblat, I. Verzhbitskiy, R. Li, Y. Cai, X. Chi, W. Fu, P. Song, A. Rusydi, G. Eda, S.A. Maier, and K.P. Loh, *Nat. Commun.* 11 (1) (2020).

[24] Y. Saeed, B. Amin, H. Khalil, F. Rehman, H. Ali, M.I. Khan, A. Mahmood, and M. Shafiq, *RSC Adv.* 10 (30), 17444 (2020).

[25] H. Tang, Y. Xu, X. Hu, Q. Hu, T. Chen, W. Jiang, L. Wang, and W. Jiang, *Adv. Sci.* 8, 2004118 (2021).

[26] X. Yang, W. Wang, R. Ran, W. Zhou, and Z. Shao, *Energy Fuels* 34 (9), 10513 (2020).

[27] F. Ji, J. Klarbring, F. Wang, W. Ning, L. Wang, C. Yin, J.S.M. Figueroa, C.K. Christensen, M. Etter, T. Ederth, L. Sun, S.I. Simak, I.A. Abrikosov, and F. Gao, *Angew. Chem.-Int. Edit.* 59 (35), 15191 (2020).

[28] J. Yin, P. Maity, M. De Bastiani, I. Dursun, O.M. Bakr, J.L. Bredas, and O.F. Mohammed, *Sci. Adv.* 3 (12), 8 (2017).

Chapter 2

Synthesis of Metal Halide Perovskites

This chapter covers the growth of bulk metal halide perovskites (MHP) single crystal (SCs), preparation of MHP films, and synthesis of MHP nanocrystals (NCs) or quantum dots (QDs). The most common approaches used for growing bulk perovskite SCs (PSCs) are based on solution or solid reactions. While the mainstream methods for preparing the MHP films are vapor deposition and solution deposition, the synthesis of MHP NCs typically involves precipitation and injection methods. Other related methods are also covered and illustrated with examples.

2.1 Bulk Metal Halide Perovskites Single Crystal Growth

2.1.1 *Background*

The synthesis of PSCs has become crucial to the study of their properties and applications. The quality of the PSCs depends strongly on the method of synthesis. In this section, we introduce the mainstream synthesis methods for growing bulk PSCs including solution and solid-phase methods. With the solution method, the PSCs are grown by changing the perovskite solubility in the precursor solution, while with the solid-phase method, the PSCs are obtained by melting the precursors at high temperature and recrystallization. We discuss the characteristics of these methods and the influence of the growth parameters on the quality and size of the resulting PSCs.

2.1.2 *Physical Basis of Perovskites Single Crystal Growth*

The basic principle of the solution method is to dissolve the perovskite precursor in a specific solvent, and then achieve the saturated state of the solution by different strategies, so as to realize the PSC growth [1]. During the progress of the solution synthesis, the perovskite precursor solution goes through three stages, namely the stable, metastable, and unstable regions. Although the concentration of the metastable region is high, the PSCs do not crystallize spontaneously. It is necessary to introduce seed crystals to precipitate PSCs. In the absence of external intervention, the concentration of the solution needs to be continuously increased until a crystal nucleus spontaneously appears inside the solution in the supersaturated region, leading to the growth of polycrystals. Therefore, the saturated region is the most favorable for the growth of large-sized PSCs. The quality of the PSCs obtained by the solid-phase method is affected by the shape of the solid–liquid interface. The solid–liquid interface usually has three shapes: convex, flat, and concave shapes. The concave shape is unfavorable for PSC growth. Once a new nucleus is formed on the tube wall, the PSC will grow in the direction perpendicular to the heat flow to form polycrystals. While in the convex shape condition, if a new nucleus appears, it will be melted quickly, favorable for the growth of large-sized PSCs. The shape of the interface can be controlled by modifying the heating and cooling strategies. In the following section, typical examples of PSC synthesis are introduced in detail [1,2].

2.1.3 *Solution Methods*

The basic principle of the solution methods is the reduction of the perovskite solubility in the precursor solution by changing the growth temperature, reducing the solvent volume, or adding anti-solvent to induce PSC crystallization. The solution methods are the most widely used for growing MHP crystals. The driving force of the seeded growth from a solution is the supersaturation state of the solute in the solvent, which can be achieved by changing the temperature or by solvent evaporation. For example, a controlled solvent evaporation technique has been developed for low temperature such as $23 \pm 0.5°C$. During the process, as the solvent evaporates at a constant rate at the fixed temperature, the solution becomes oversaturated gradually, and the nucleation seed is formed and further growth into a larger crystal follows. Once the nucleation seed is formed, it is desirable to control the solution conditions so that extraneous nucleation is terminated and further crystal growth proceeds only on the existing seed.

The overall crystallization process can be separated into two steps, nucleation and growth. The basic physics of nucleation can be explained using a classical nucleation theory in which the nucleation rate, j_0, can be expressed by the Arrhenius-type equation [3]:

$$j_0 = K \exp\left(-\frac{\Delta G^*}{k_B T}\right) \qquad (2.1)$$

where K is a constant closely associated with oversaturation, ΔG^*, the Gibbs free energy change for the nucleus formation, k_B, the Boltzmann constant, and T, the temperature. According to the equation, there is a critical supersaturation level below which the nucleation rate is practically zero and beyond which it increases sharply. The reason for this behavior is that the pre-exponential term K in the equation is a very weak function of supersaturation compared with the exponential term $\exp\left(-\frac{\Delta G^*}{k_B T}\right)$, which varies strongly with temperature. Therefore, to suppress extraneous nucleation and achieve growth of only the desired high-quality SCs, the growth conditions can be controlled by regulating the temperature. To obtain the best SCs, the process is controlled to maximize the crystal growth and minimize ancillary nucleation. In a solution system, the crystal growth rate is determined by two factors: (1) the deposition rate of the solute molecule onto the crystal surface and (2) the diffusion rate of the solute in the solution. If the solute deposition rate is slower than its diffusion rate, the former is the limiting factor for the crystal growth, and the oversupplied solute will lead to the formation of defective structures, including twins and fine crystallites. Otherwise, the solute diffusion limits the crystal growth rate, which is the desired situation for growing high quality SCs. The diffusion rate increases exponentially with temperature, as shown in Eq. (2.2) [4]

$$D = D_0 = \exp\left(-\frac{\Delta H}{RT}\right) \qquad (2.2)$$

where D_0 is the diffusion constant, and ΔH is the diffusion activation energy. Too high a solute diffusion rate will result in imperfections in the SCs. Hence, it is critical to control the temperature gradient or ramp rate to keep the growth conditions within the supersaturated zone, separated by the solubility curve and nucleation curve. Meanwhile, to attain the best SC quality, it is equally important to use low temperature (<60°C) to achieve the optimal solute diffusion rate while accomplishing substantial crystallization yield.

2.1.3.1 *Controlled evaporation growth (CEG)*

In the controlled evaporation growth (CEG) method (or low temperature solvent volatilized crystallization method), the perovskite precursors are dissolved in an organic or acid solvent at a concentration close to the saturated state and then the solvent is evaporated from the multiple holes on a container lid at a controllable rate. High-quality PSCs can be synthesized after a long time of solvent evaporation at a certain temperature. The CEG method is suitable for growing both 2D and 3D PSCs. For example, Zhang *et al.* prepared 2D Ruddelsden–Popper (RP) phase (Phenethylammonium)$_2$PbBr$_4$ (PEA$_2$PbBr$_4$) PSCs by dissolving PEABr and PbBr$_2$ materials in *N,N*-dimethylformamide (DMF) [4]. To control the evaporation rate during the growth process, an array of 2×2 holes of 1.0 mm diameter was pre-made in the sealing film of the glass container. An acid solution can also be utilized as an evaporation solvent. Raghavan *et al.* prepared (*n*-butylammonium)$_2$(methylammonium)$_{n-1}$Pb$_n$I$_{3n+1}$ (BA$_2$MA$_{n-1}$PbnI$_{3n+1}$) (n = 1, 2, and 3) PSCs using hydroiodic acid (HI) as the solvent, as shown in Figures 2.1(a)–2.1(c) [5]. The precursor solution was kept at 60°C and slowly evaporated, and 2D BA$_2$MA$_{n-1}$PbnI$_{3n+1}$ RP phase PSCs were obtained within 1 day. Similarly, Nandi *et al.* have demonstrated 3D MAPbCl$_3$ bulk PSCs grown by CEG method. Using a mixture of dimethyl sulfoxide (DMSO) and GBL as the solvent in which lead (II) chloride (PbCl$_2$) and methylammonium chloride (MACl) were dissolved, millimeter-sized MAPbCl$_3$ PSCs were obtained by solvent evaporation without any disturbance at room temperature after four weeks [6]. Ye *et al.* greatly shortened the crystal growth cycle by

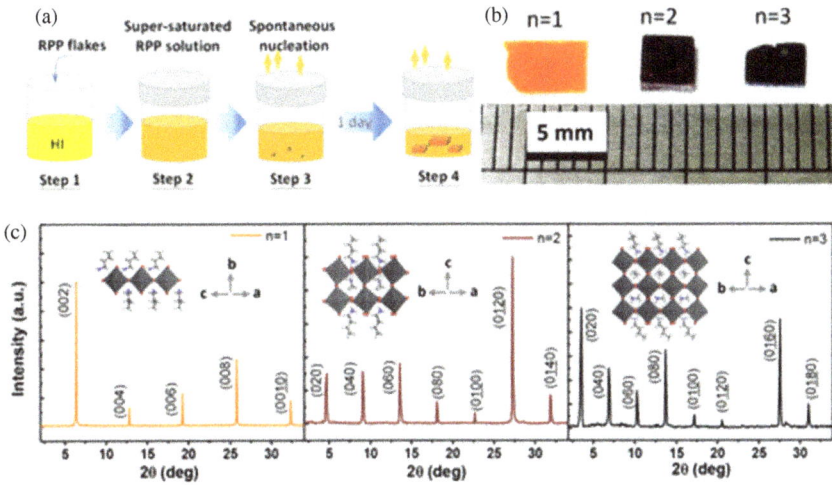

Figure 2.1. (a) Schematic diagram of the growth process of homologous 2D RPP PSCs using the CEG method. (b) Photograph of as-grown crystals of $n = 1$, 2, and 3. (c) X-ray diffraction patterns of the grown crystals. The insets in panel c represent the corresponding crystal structures with different perovskite layer thicknesses. Reproduced from Ref. [5].

utilizing the flowing gas to accelerate the solvent evaporation process. A homogenous saturated solution was first prepared by continuous stirring at 60°C (except for X = Br at room temperature) for 2 hours. Using DMF as the solvent, they heated the perovskite precursor solution to 120°C and placed it in a flowing nitrogen gas atmosphere. A centimeter-scale MAPbI$_3$ PSC was finally obtained within 2–3 h, MAPbBr$_3$ PSC was finally obtained within 12 h, and MAPbCl$_3$ PSC was finally obtained within 5 days [7]. With the increase of evaporation temperature or gas flow rate, solvent evaporation is accelerated, resulting in the greatly shortened growth time of PSCs.

2.1.3.2 *Liquid-diffused separation-induced crystallization*

Yao *et al.* developed a room-temperature liquid-diffused separation-induced crystallization (LDSC) that uses silicone oil to separate the solvent from the perovskite precursors for growing high-quality PSCs, as shown in Figure 2.2 [2]. This method is similar to the CEG method in

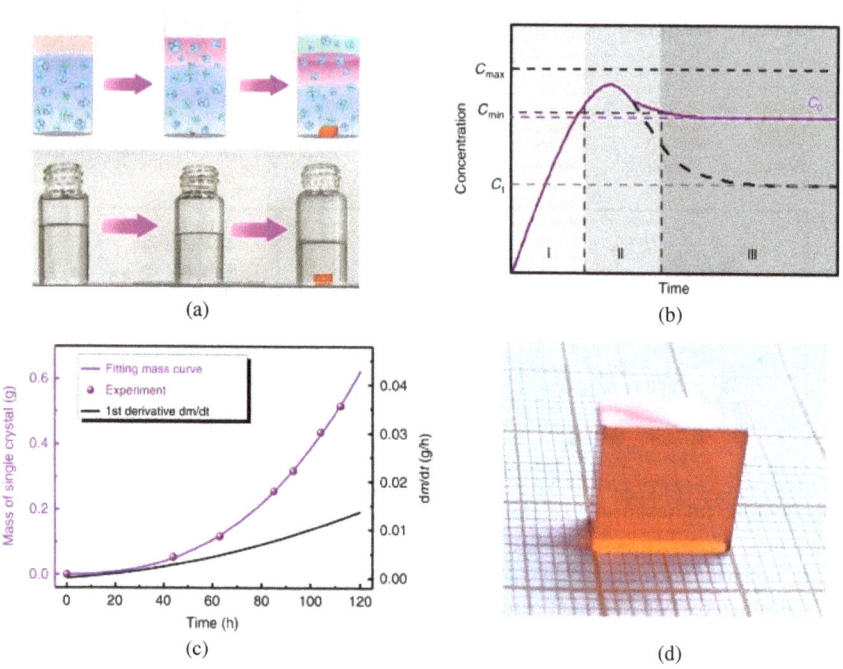

(a)

(b)

(c)

(d)

Figure 2.2.   (a) Schematic illustration of growth diagram of perovskite PSCs using the RT LDSC method. (b) Schematic representation of the concentration of solution before and after nucleation as a function of time. C_t is the solubility, C_0 is the concentration of perovskite solution ($C_0 > C_t$), C_{min} is the minimum concentration for nucleation, C_{max} is the maximum concentration for nucleation. The regions I, II, and III represent prenucleation, nucleation, and growth stage, respectively. (c) The mass and mass derivative as a function of the growth time for the MAPbBr$_3$ PSCs. (d) Photograph of MAPbBr$_3$ PSCs grown using the RT LDSC method. Reproduced from Ref. [2].

terms of solvent reduction to obtain supersaturated solutions. The perovskite solution and silicone oil were injected into the bottle successively. As the solvent and silicone oil have different densities and polarities, the solvent diffuses into the silicone oil and the precursor solution becomes more concentrated, leading to the formation of an oversaturated perovskite precursor solution. Bulk high-quality 3D MAPbX$_3$ (X = Cl, Br, and I) SCs of $2.0 \times 2.0 \times 1.0$ mm^3 size were obtained using this method. Compared with PSCs obtained under heating conditions, room temperature growth eliminates the influence of thermal gradient convection in the growth process, and thus avoids winding defects and cracks in the resulting PSCs.

2.1.3.3 *Anti-solvent vapor-assisted crystallization (AVC) method*

In the anti-solvent vapor-assisted crystallization (AVC) method, the growth of PSCs is realized based on the different solubilities of perovskite compounds in different solvents. The solvent with high solubility and moderate coordination is called a good solvent and that with low solubility is called a bad solvent (or anti-solvent) [8,9]. The schematic diagram of the crystallization process is presented in Figure 2.3. First, the perovskite precursors are dissolved in a good solvent in a small bottle. Subsequently, the bottle is placed in a large container with a bad solvent, and the container is then sealed. As the bad solvent gradually volatilizes into the perovskite precursor solution, the solubility of the mixed solution decreases, which results in the growth of PSCs at the bottom of the small bottle. In this method, dimethyl formamide (DMF), dimethylsulfoxide (DMSO), and γ-butyrolactone (GBL) are commonly

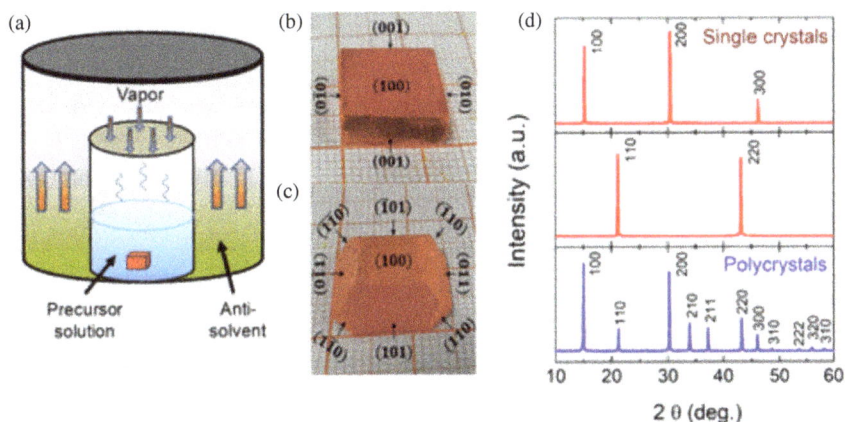

Figure 2.3.   (a) Schematic diagram of MAPbBr$_3$ crystals grown from the solution using the AVC method. (b, c) The photograph of as-grown cubic and polyhedral MAPbBr$_3$ crystals with the mixed anti-solvent of DCM/DMF = 1:2. (d) {100} and {110} facets are identified by the single-crystal XRD patterns. The powder XRD pattern of MAPbBr$_3$ shows the space group Pm3m. Reprinted from Ref. [8].

used as good solvents while chlorobenzene, chloroform, benzene, xylene, 2-propanol, toluene, and acetonitrile are often used as anti-solvents during the growth of metal halides. However, choosing the right solvent/anti-solvents is key to crystal growth. Therefore, the choice of anti-solvent is very important, which directly affects the quality and size of PSCs. Shi *et al.* employed the AVC method to prepare sizable MAPbX$_3$ (X = Br and I) 3D PSCs with volumes exceeding 100 mm^3 crack-free smooth surfaces and distinguishable facets using DMF or GBL as the good solvent and dichloromethane (DCM) as the bad solvent [10]. Then, Ge *et al.* prepared centimeter-sized MAPbCl$_3$ and MAPbBr$_3$ PSCs using a mixed solution of DMSO and DMF as the good solvent and DCM as the bad solvent [11]. Millimeter-sized all-inorganic CsPbBr$_3$ PSCs synthesized using the AVC method have also been reported. Ding *et al.* selected DMSO as the good solvent and a mixed solution of methanol (MeOH) and ethanol as the anti-solvent, by controlling anti-solvent vapor diffusion velocity and growth temperature to grow CsPbBr$_3$ PSCs [12]. The optimized growth temperature was found to be 40°C. Rakita *et al.* utilized DMSO as the good solvent and acetonitrile (MeCN) or MeOH as the anti-solvent to grow high-quality CsPbBr$_3$ PSCs [13]. Before sealing, a precursor solution was first titrated with MeCN or MeOH to reach the saturated state, which was important to prevent precipitation of the undesired byproducts of Cs$_4$PbBr$_6$ or CsPb$_2$Br$_5$. Similarly, Zhang *et al.* obtained centimeter-sized CsPbBr$_3$ crystals by a modified AVC solution method at low temperature. To improve the size and crystalline quality of CsPbBr$_3$ crystals, they dilute the MeOH solution with DMSO to tailor the growth speed. The vapor pressure of the anti-solvent MeOH is controlled by changing the molar ratio of MeOH and DMSO in the anti-solvent. At the ratio of 50%, a CsPbBr$_3$ crystal with a size of $42 \times 5 \times 3$ mm^3 was obtained [14].

2.1.3.4 *Inverse temperature crystallization method*

The inverse temperature crystallization (ITC) method is mainly used for growing PSCs in a certain solvent. The main feature of this method is the decrease in perovskite solubility with the increase in temperature. Large-sized PSCs can be obtained by gradually increasing the temperature. The optimal solvents for the ITC method were reported to be DMF, DMSO, GBL, or their mixtures. To obtain large-sized and high-quality PSCs, various synthesis schemes were introduced, resulting in the growth of a series of PSCs, such as MAPbX$_3$ (X = Cl, Br, and I), HC(NH$_2$)$_2$PbX$_3$ (FAPbX$_3$) (X = I and Br), and CsPbBr$_3$. Bakr's group first found the inverse temperature-solubility behavior of MAPbX$_3$ perovskites in certain solvents [15]. A schematic diagram of the ITC method is shown in Figure 2.4(a). This novel phenomenon in hybrid perovskites enabled them to design an innovative crystallization method for these materials to rapidly grow high-quality size- and shape-controlled SCs of both MAPbBr$_3$ and MAPbI$_3$. To accelerate the preparation of PSCs, they further optimized the ITC method for growing MAPbX$_3$ (X = I or Br) PSCs by carefully selecting the solvent (DMF for MAPbBr$_3$; GBL for MAPbI$_3$) and heating temperature (80°C for MAPbBr$_3$ and 110°C for MAPbI$_3$), and high-quality millimeter-sized MAPbBr$_3$ PSCs can be achieved within 3 hours, as shown in Figure 2.4(b).

Likewise, Kadro *et al.* used the ITC method for the growth of free-standing crystals of CH$_3$NH$_3$PbI$_3$ with 1 mm length within minutes from solution without the addition of any capping agents or seed particles [16]. They used PbI$_2$ and CH$_3$NH$_3$I as precursors and GBL as solvent. Both salts in a molar ratio of 1:1 were dissolved under vigorous stirring in GBL at 100°C until clear solutions were obtained. The temperature was rapidly increased to 190°C and maintained at that temperature until the formation of desired dimensions. For concentrations stable at room temperature, crystals start to form when the solution temperature exceeds 135°C. Large crystals of 1 mm in length could be grown within minutes at precursor concentrations of 0.905 M CH$_3$NH$_3$PbI$_3$. Ding *et al.* further used the ITC

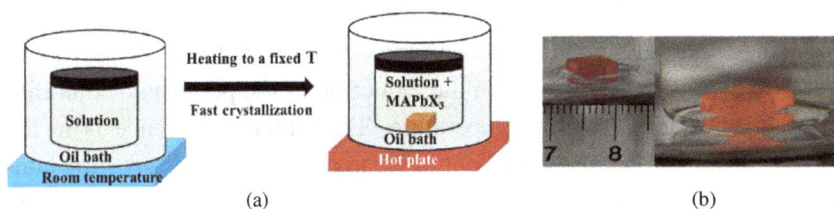

Figure 2.4.   (a) Schematic representation of the ITC apparatus in which the crystallization vial is immersed within a heating bath. (b) MAPbBr$_3$ crystal grown by ITC method. Reproduced from Ref. [15].

method to synthesize MAPbCl$_3$ and MAPbBr$_3$ bulk PSCs [17]. To grow MAPbBr$_3$ PSC, PbBr$_2$, and MABr in a molar ratio of 1:1 were dissolved in DMF and stirred until a transparent solution was obtained. A seed crystal with a size of 1 mm was added into the solution, which was maintained at 80°C for 2 days. A large-sized MAPbBr$_3$ SC was obtained. Similarly, MAPbCl$_3$ SCs were grown from a mixed solution of DMF and DMSO (1:1 in volume) with dissolved PbCl$_2$ and MACl in a molar ratio of 1:1. At the initial stage, the transparent solution was sealed and maintained at 80°C for 7 days. When the solution became creamy in color, a small cubic crystal was added to the solution. A large MAPbCl$_3$ SCs could grow within 3 days.

In addition, Saidaminov *et al.* successfully synthesized the target millimeter-sized FAPbX$_3$ (X = I or Br) PSCs using the ITC method [18]. To grow FAPbI$_3$, FAI and PbI$_2$ were dissolved in GBL at 80°C; then, the solution was heated to 115°C to obtain the FAPbI$_3$ PSCs. Subsequently, by increasing the temperature of the solution to 117°C or 120°C, the PSC size can be further increased. While for the growth of FAPbBr$_3$ PSCs, the solvent was changed to a mixed solution of DMF and GBL (1:1 V/V), and the optimized growth temperature was set at 55°C. The ITC method is also suitable for the growth of all-inorganic PSCs. By controlling the molar ratio of perovskite precursors, Saidaminov *et al.* obtained pure CsPbBr$_3$ bulk PSCs [19]. The CsBr and PbBr$_2$ with a molar ratio of 1:2 were dissolved in DMSO at 60°C , and the temperature of the solution was slowly increased to 120°C at a rate of 2.5°C h^{-1}. After 10 hours, a millimeter-sized CsPbBr$_3$ PSC was obtained.

2.1.3.5 *Low-temperature-gradient crystallization method*

Similar to the ITC method, the low-temperature-gradient crystallization (LTGC) method is also based on the reduction of the perovskite solubility by increasing the solution temperature. Their main difference is in the control of the temperature gradient or ramp rate to keep the growing conditions within the supersaturated zone at a low temperature (<60°C) to achieve optimal solute diffusion rate while accomplishing substantial crystallization yield. For example, Liu *et al.* report an LTGC method for high-quality MAPbBr$_3$ PSC with lateral dimension as large as two inches [20,21]. They found that when the temperature increased from 20 to 60°C, the solubility of MAPbBr$_3$ in DMF significantly decreased, from 0.89 to 0.47 g·mL^{-1}. However, when the temperature continued to increase to 100°C, the solubility exhibited a much smaller change, decreasing from 0.47 to 0.29 g·mL^{-1}. Based on this, the turning temperature was set to 60°C. For the growth of MAPbBr$_3$ PSC, MABr and PbBr$_2$ were dissolved, with a 1:1 molar ratio, in DMF solvent at room temperature and stirred for 24 hours. A transparent saturated solution was produced by filtrating the mixed solution. When the solution was slowly heated up from 25 to 60°C at the heating rate of 2°C d^{-1}, crystals were slowly formed. Finally, a 44 × 49 × 17 mm large SC was obtained after 18 days. The entire process of PSC growth and the photograph of MAPbBr$_3$ PSCs are presented in Figure 2.5. Yuan *et al.* synthesized large-sized MAPbCl$_3$ PSCs using the LTGC method [22]. For the growth of CH$_3$NH$_3$PbCl$_3$ PSCs, MACl and PbCl$_2$ were mixed in a 1:1 molar ratio, and dissolved in a mixed solution of DMF and DMSO (the volume ratio was 1:1). Various PSC growth temperatures were attempted, e.g., 60, 70, 80, 100, and 120°C, and the results revealed that 70°C was the optimal temperature for PSC growth, at which a PSC with an ideal size of 6 × 6 × 2 mm^3 was obtained.

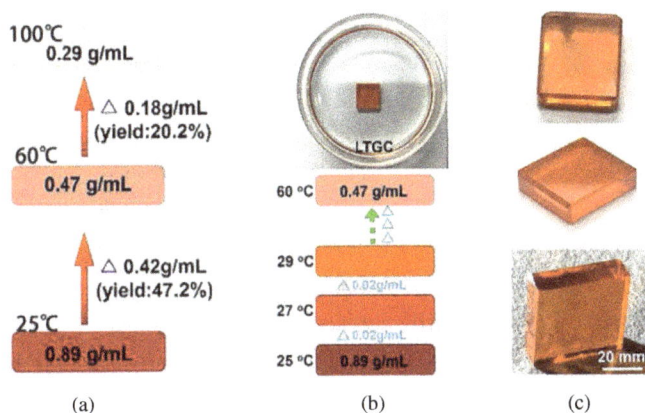

Figure 2.5. (a) Schematic diagram of the crystallization process of MAPbBr$_3$ PSCs using the LTGC method. (b) Photographs of MAPbBr$_3$ PSCs. Reproduced from Ref. [20].

2.1.3.6 *Solution temperature-lowering crystallization method*

The mechanism of the solution temperature-lowering crystallization (STLC) method is opposite to that of the ITC method, i.e., the reduction of the perovskite solubility in the precursor solutions mainly by lowering the temperature, with the acid solution generally used as the solvent. For instance, in 2015, Dang *et al.* first reported the growth of tetragonal MAPbI$_3$ PSC with dimensions of 10 mm × 10 mm × 8 mm using an STLC method in HI solution [23]. For the growth of MAPbI$_3$ PSC, Pb[(CH$_3$COOH)$_2$]·3H$_2$O (37.933 g, 0.100 mol) was dissolved in aqueous HI (260 ml), under constant stirring, forming a yellow solution. Then, newly synthesized MAI (15.9 g, 0.100 mol) was added to the yellow solution. With the decrease of the solution temperature from 65 to 40°C to induce the saturation of the solute, the black and shiny crystal of MAPbI$_3$, with the size of 10 mm × 10 mm × 8 mm, was successfully grown after several days and was washed and filtered with HI and then acetone. After that, the growth of bulk PSCs by this method has been largely reported. This method is also suitable for the growth of mixed-halide PSCs. For example, Fang *et al.* prepared the mixed-halide MAPbBr$_{3-x}$Cl$_x$ PSCs in the mixed solutions of hydrobromic acid (HBr) and hydrochloric acid (HCl), and MAPbI$_{3-x}$Br$_x$ PSCs in the mixed solutions of HBr and HI acids with different molar ratios. The precursor solution was prepared by mixing methylamine, single or mixed haloid acid with different halide ratios, and lead (II) acetate to form a supersaturated aqueous solution at 100°C. The SCs with a thickness of 0.5–5 mm were obtained from the precursor solution by gradually lowering the temperature from 100°C to room temperature [24].

For the growth of 2D PSCs, Li *et al.* selected HI as the solvent for dissolving BABr, MABr, and PbBr$_2$ [25]. Subsequently, the solution temperature was slowly decreased from 80°C to room temperature at a rate of 0.5°C d^{-1}. After 7 days, millimeter-sized 2D RP phase BA$_2$MA$_2$Pb$_3$Br$_{10}$ multi-layered hybrid PSCs were obtained. Furthermore, 2D RP phase BA$_2$MA$_{n-1}$PbnBr$_{3n+1}$ and BA$_2$MA$_{n-1}$PbnI$_{3n+1}$ (n = 1–5) PSCs were obtained by this method, as shown in Figure 2.6 [26] . By using the STLC method, Chen *et al.* also successfully prepared 2D RP phase BA$_2$Cs$_{n-1}$PbnBr$_{3n+1}$ (n = 1–3) PSCs [27]. To further increase the size of 2D RP phase PSCs,

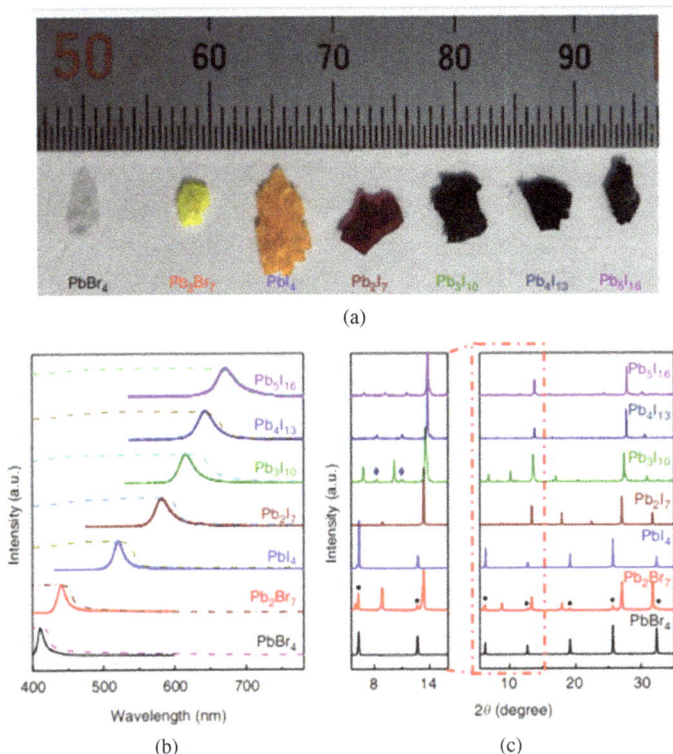

Figure 2.6. Characterizations of 2D PSCs using the STLC method. (a) Photograph of the as-synthesized PSCs plates. From left to right: $(BA)_2(MA)_{n-1}PbnBr_{3n+1}$ and $(BA)_2(MA)_{n-1}PbnI_{3n+1}$. (b) Absorption (dash line) and photoluminescence (solid line) spectra of the perovskite plates. (c) Powder XRD spectra of the perovskite crystal plates. The low angle part is magnified to see the diffraction peaks for impurity phases. Black dot: $(BA)_2PbBr_4$ phase within $(BA)_2(MA)Pb_2Br_7$; Navy rhombus: $(BA)_2(MA)_3Pb4I_{13}$ phase within $(BA)_2(MA)_2Pb_3I_{10}$. Reproduced from Ref. [26].

Liu *et al.* developed a modified STLC method for growing high-quality large-sized PEA_2PbBr_4 PSCs in GBL [28].

2.1.3.7 *Acid-catalyzed hydrolysis growth method*

The synthesis of $MAPbX_3$ (X = Cl, Br, and I) PSCs is generally believed to have started from expensive MAX salt precursors [29]. For this, alternative acid-catalyzed hydrolysis growth (ACHG) method was developed.

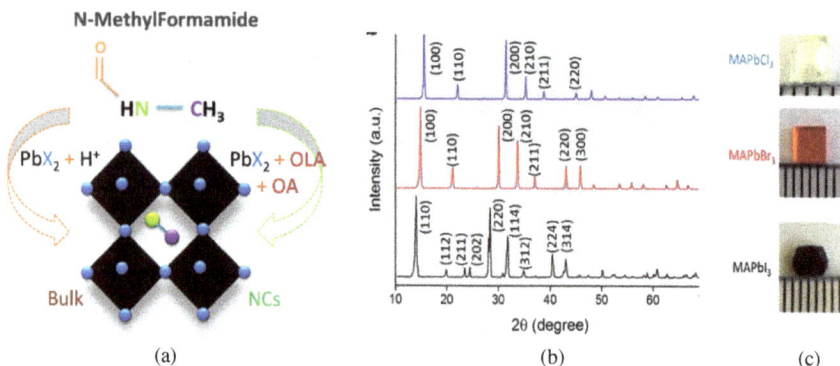

Figure 2.7.   (a) Sketch of MAPbX$_3$ PSCs obtained using the ACHG method employing NMF as a source of MA. (b) Indexed XRD patterns of ground PSCs. (c) Photographs of MAPbX$_3$ PSCs. Reproduced from Ref. [29].

Shamsi *et al.* successfully obtained MAPbX$_3$ PSCs from *N*-methylformamide (NMF) and PbX$_2$ in the presence of hydrohalic acid (HX) via acid-catalyzed hydrolysis of NMF. The concentrations of Pb^{2+} and acid and the ratio of acid to NMF in the precursor solution were optimized, as presented in Figures 2.7(a) and 2.7(b). In this method, MAX salts were creatively replaced with a less-toxic NMF as the source of MA$^+$. Eventually, millimeter-sized MAPbX$_3$ PSCs were obtained. The shapes of the PSCs are presented in Figure 2.7(c), which are consistent with the previously reported MAPbCl$_3$ [30], FAPbBr$_3$ [8,31], and MAPbI$_3$ PSCs [32,33] synthesized using ammonium salts.

2.1.3.8 *Seed solution growth method*

The seed solution growth method is mainly used for growing large-sized PSCs through the selection of high-quality pre-synthesized small PSCs as seeds. According to the different seed positions in the solution, the method can be divided into top- and bottom-seeded solution growth. For example, Lian *et al.* designed a special setup to grow centimeter-sized MAPbI$_3$ PSCs using the bottom-seeded solution growth (BSSG) method, as shown in Figures 2.8(a) and 2.8(b) [34]. A carefully selected seed crystal produced by the STLC method was introduced to the freshly perovskite-precursor-saturated solution at a temperature of 100°C. Pt

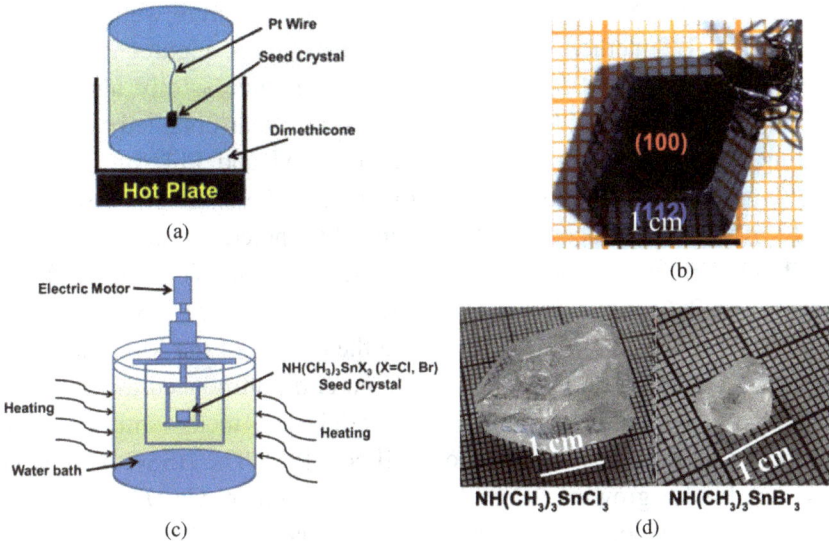

Figure 2.8. Bulk PSC grown using the BSSG method. (a) Schematic diagram of the Pt wire-assisted BSSG growth setup, and (b) photograph of a bulk MAPbI$_3$ PSC. Reproduced from Ref. [34]. (c) Schematic diagram of the electric rotating setup, and (d) photographs of NH(CH$_3$)$_3$SnX$_3$ PSCs. Reproduced from Ref. [35].

wire was used to fix the seed PSC near the bottom of the flask, preventing contact with the flask. The temperature was kept at 100°C for 10 min to dissolve the outer surface of the seed crystal, then it quickly decreased to 82°C and was followed by a slow cooling rate until the desired size was obtained. Subsequently, it dropped to 57°C at a slow cooling rate of 5 K/h. Finally, the desired centimeter-sized MAPbI$_3$ PSCs was taken out from the growth solution at a temperature higher than 47°C.

Similarly, Tin-based PSCs have also been synthesized in an ambient atmosphere by the BSSG method. Dang *et al.* synthesized large NH(CH$_3$)$_3$SnX$_3$ (X = Cl or Br) PSCs under water bath heating [35]. A suitable selected seed crystal was obtained via the STLC method and placed on a plate and kept rotating at a constant speed during the growth process. The slow rotation helps the uniform mixing of the solution and reduces the radial temperature gradient to prevent the components from supercooling. Finally, with the decrease in temperature from 56°C and 54°C to room temperature, an NH(CH$_3$)$_3$SnBr$_3$ PSC with a size of 13 × 8 × 6 mm^3

and an $NH(CH_3)_3SnCl_3$ PSC with a size of $8 \times 6 \times 4$ mm^3 were obtained using the STLC method, respectively, as presented in Figures 2.8(c) and 2.8(d). By using the BSSG method, Liu *et al.* successfully obtained a 2-inch-sized MAPbI$_3$ PSC [36,37]. The seed PSC was obtained by slowly heating the perovskite precursor solution (MAI and PbI$_2$ in GBL) to 100°C and kept at this temperature overnight. The small-sized selected seed PSC was introduced to a fresh perovskite-precursor-saturated solution to perform a new round of PSC growth using the ITC method. After 7 rounds of PSC growth, a MAPbI$_3$ PSC with a size of $71 \times 54 \times 39$ mm^3 was finally obtained. To further improve the quality of the MAPbI$_3$ PSCs and enhance the crystallization speed, Lian *et al.* introduced the chlorine element by mixing methylamine hydroiodide and methylamine hydrochloride with the same molar ratio in HI acid [38]. The BSSG method is also suitable for growing 2D bulk PSCs. For example, Ge *et al.* successfully obtained the desired centimeter-sized 2-phenylethylammonium lead bromide (PEA$_2$PbBr$_4$) PSCs using the BSSG method [39].

Similar to the BSSG method, Dang *et al.* successfully obtained centimeter-sized MASnI$_3$ and FASnI$_3$ PSCs using the top-seeded solution growth (TSSG) method based on the electric rotating setup, where the seed PSCs were located on the upper side of the solution [40]. SnO and MAI were dissolved in a mixed acid solution of HI and HPO$_2$ at 75°C to form a perovskite precursor solution, and then the seed PSC synthesized using the STLC method was put on the platform of the setup, which was rotated at a constant speed as the temperature decreased. After maintaining for a month, large-sized MASnI$_3$ and FASnI$_3$ PSCs were finally obtained using the STLC method. Furthermore, Dang *et al.* successfully synthesized a centimeter-sized Cs$_2$AgBiBr$_6$ PSCs by a TSSG method, as presented in Figure 2.9 [41]. The CsCO$_3$, Ag$_2$CO$_3$, and (BiO)$_2$CO$_3$ precursors were dissolved in HI and H$_3$PO$_2$ mixed solution at 80°C with constant stirring to form a transparent solution. Then solutions were saturated at 75°C. By slowly decreasing temperature, large-sized Cs$_2$AgBiBr$_6$ PSCs were grown using TSSG method in an ambient atmosphere after 2 weeks. Dong *et al.* established a new setup to grow large-sized MAPbI$_3$ PSCs using the TSSG method [42]. The seed PSC was initially obtained using the STLC method. Subsequently, the selected high-quality seed placed on a silicon slice was introduced to the growth solution in a sealed bottle at a

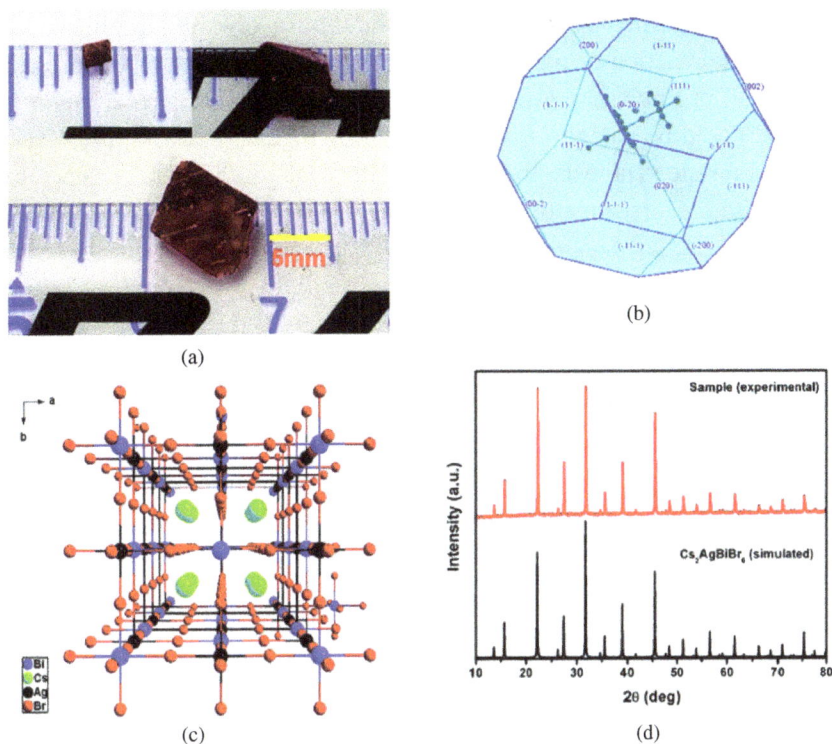

Figure 2.9. Photographs of $Cs_2AgBiBr_6$ PSCs growth using the TSSG method. (b) Theoretical morphology of the $Cs_2AgBiBr_6$ PSCs deduced by the Bravais–Friedel–Donnay–Harker (BFDH) method. (c) Crystal structure of $Cs_2AgBiBr_6$ at room temperature. (d) Comparison between the experimental and simulated X-ray diffraction patterns of $Cs_2AgBiBr_6$. Reproduced from Ref. [41].

constant temperature of 75°C. By creating a temperature gradient between the upper and lower parts of the solution via partial air cooling, a bulk $MAPbI_3$ PSC with a size of 10 mm in length and 3.3 mm in height was obtained.

2.1.4 *Solid-Phase Method*

The solution methods are simple and convenient and are also suitable for growing a variety of PSCs. However, finding suitable solvents that exhibit optimized volatility or solubility is difficult. To solve this problem, the solid-phase reaction of perovskite precursors under high temperature can be used. Until now, the growth of PSCs using the solid-phase (melting) method could be categorized into vertical Bridgman and simplified vacuum solid state synthesis. In both cases, the desired PSCs are obtained by perovskite precursor melting and cooling crystallization.

2.1.4.1 *Vertical Bridgman growth method*

The growth of PSCs by the vertical Bridgman growth (VBG) method usually requires a specific Bridgman furnace. In the process, the quartz ampoule was filled with all of the raw reactants, and then sealed under vacuum, and moved down through the temperature gradient. At a definite temperature, spontaneous nucleation was induced at the tip of the ampoule, and the crystallization front propagated through the molten material [43,44]. The Bridgman growths of $CsSnI_3$ poly-crystalline ingots and $CsPbBr_3$ SCs were prepared by the Kanatzidis group [45,46]. Compared with the solution method, the PSCs obtained using the VBG method are much larger and have a higher quality. This is attributed to that fact that as the whole process is under water- and oxygen-free conditions, unnecessary contamination can be avoided [46,47]. Recently, the preparation of large-sized inorganic $CsPbBr_3$ PSCs using this method has attracted much attention. Zhang *et al.* used a simplified VBG method to grow $CsPbBr_3$ PSCs in a home-made vertical two-zone furnace [48]. First, $CsPbBr_3$ polycrystals were obtained via chemical co-precipitation [49]. Then polycrystalline powders were sealed under a 0.1 Pa pressure in a fused silica ampule which had a conical tip. The tube furnace was first heated up to the melting point of $CsPbBr_3$ polycrystalline powders and kept at this temperature for a few hours. Then the tube furnace was cooled for crystal growth. The cooling process mainly consisted of three steps. First, the cooling process was decreased at a rate of 10°C/h until the minimum temperature of the ingots reached 130°C. Then, the temperature was very slowly decreased at rate of 1°C/h until the temperature of the whole ingot was below 130°C. Finally, the temperature of the furnace was decreased to room temperature in few hours to obtain a $CsPbBr_3$ PSC with a length of about 60 mm. Through the precise control of the cooling rate, the formation of mixed and impurity phases can be effectively avoided [50]. Song *et al.* prepared ultra-large $CsPbBr_3$ SCs with a volume of 25 cm^3 using a modified VBG method [51]. A $CsPbBr_3$ seed was first synthesized by mixing CsBr and $PbBr_2$ powders to Silica ampoule in a rocking furnace to achieve a sufficient and homogeneous reaction. The ampule was then removed from the four-zone Bridgman furnace for PSC growth. Through the optimization of the melting temperature and cooling strategies, He *et al.* obtained a transparent and large $CsPbBr_3$ PSC ingot free of visible cracks or grain boundaries, which was further cut into wafers with different sizes, as shown in Figure 2.10 [19,52]. By controlling the melting temperature, temperature gradient, cooling strategies, and other parameters in the VBG method, ultra-large PSCs can be obtained. However, this

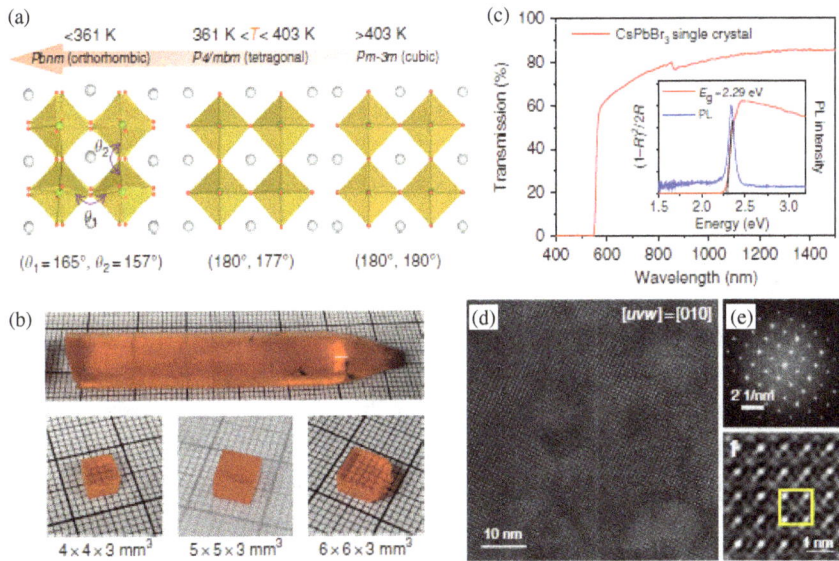

Figure 2.10. Crystal growth and general properties of $CsPbBr_3$. (a) Non-destructive phase transitions in $CsPbBr_3$ during cooling involving small rotations and tilts of $PbBr_6$ octahedral. (b) As-grown SC ingot with a diameter of 11 mm, and the SC wafers with different sizes. The pure orthorhombic phase of the crystal was confirmed by powder X-ray diffraction. (c) Optical transmission spectrum for $CsPbBr_3$ SCs with a size of $5 \times 5 \times 3$ mm^3. Insets are the optical absorption spectrum obtained from diffuse reflectance measurement on ground $CsPbBr_3$ crystals and the steady-state PL of $CsPbBr_3$ SCs excited at 440 nm. (d) High resolution TEM (HRTEM) with selected area electron diffraction (SAD) in (e), and magnified lattice image in f. Yellow squares in (f) are drawn to compare HRTEM image. Reproduced from Ref. [52].

method requires a special Bridgman furnace, which limits the application of this method to a certain extent. Moreover, the as-prepared PSCs are in direct contact with the quartz tube, which tends to generate mechanical stress and thus affects their wide applications [52].

2.1.4.2 *Vacuum solid-state growth method*

The principle of the vacuum solid-state growth (VSSG) method for the growth of PSCs is similar to that of the VBG method. However, the VSSG method is much simpler without the need of a specific furnace. The PSC growth process mainly consists of three steps: grinding, sealing, and annealing. The precursor powders are initially ground fully and loaded into the silica tube under an inert gas environment. Then, the silica tube is vacuumed and sealed. Finally, the silica ampoule is annealed in a muffle furnace at a high temperature for a certain time until the precursors completely react; then, it is slowly cooled down to room temperature for crystallization. Via this method, a series of 2D RP phase all-inorganic mixed halide perovskite ($Cs_2PbI_2Cl_2$), indium-based charge-ordered halide perovskite ($Cs_2In(I)In(III)Cl_6$), rare-earth-based halide compounds, and Au^{3+} semiconductor ($AuPb_2I_7$) crystals have been grown [47,53–55]. The annealing temperature, time, and cooling rate in the VSSG method are all critical for obtaining PSCs with high quality and the right phase. The VSSG method simplified the crystal melting and growth process using a muffle furnace. Moreover, this method has no solubility requirement of the precursors in a specific solvent as in the solution method. The amount of the precursors required for growth is much less than that of the VBG method, which high-lights its advantages in exploring new halide perovskite materials. However, it is difficult to obtain large-sized PSCs without precise temperature control.

2.2 Metal Halide Perovskites Film Preparation

2.2.1 *Background*

The film quality directly affects the performance of perovskite thin film devices, such as light absorption, carrier diffusion length, charge transfer, and recombination. This effect is particularly prominent in planar perovskite solar cells. Poor quality perovskite films are accompanied by low light absorption and the recombination of unwanted electron holes, which greatly affect the performance of the device. Therefore, finding a suitable method to prepare perovskite films with high crystallinity, high uniformity, and low defect density is the prerequisite for obtaining devices with high stability and high conversion efficiency.

The perovskite film-forming methods can be divided into the vapor deposition method and the solution deposition method. The vapor deposition method is to sublime the raw materials for preparing perovskite materials by heating the beam source under vacuum and depositing them on the substrate to obtain perovskite thin films. The Solution deposition method refers to dissolving the raw material components for the preparation of perovskite materials in a solvent, then depositing them on the substrate by a certain method, and then removing the solvent to obtain a perovskite film. According to whether the raw materials are deposited simultaneously or in steps, the solution deposition method can be divided into one-step method and two-step method.

2.2.2 *Physical Basis of Films Preparation*

2.2.2.1 *Thin film morphologies and microstructures*

The perovskite film morphology, that is, continuous substrate coverage and low surface roughness, is of paramount importance for highly efficient solar cells and optoelectronic devices in general [56]. It is generally believed that most perovskite-type active layers need to have large grains to facilitate the movement of carriers, because small grains will bring a large number of grain boundaries parallel to the substrate to block the movement of carriers [57]. Of course this conclusion is not absolute, LED active layers usually require a small grain structure, which helps to spatially confine electron–hole pairs and enhance radiative recombination [58]. Regardless of whether the grains are large or small, however, all device active layers must generally have a highly dense microstructure consisting of closely packed grains without pinholes or other voids [59]. Controlling the chemical properties of the precursor solution is one of the

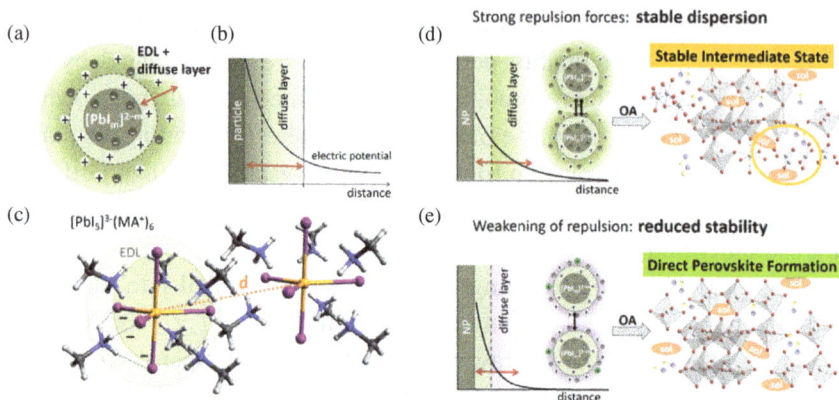

Figure 2.11. Illustration of DLVO (Derjaguin, Landau, Verwey, and Overbeek) theory applied to $MAPbI_3$. (a) Possible formation of $MAPbI_3$ particles based on the EXAFS data fit. (b) The decrease in the electric potential concerning the distance from the particle surface. (c) A hypothetical model illustration for the formation of an EDL surrounding the NP, considering the hydrogen bonds of the counter ion MA+ (gray, dotted lines). (d) Well-pronounced repulsion forces between particles enable the extraction of intermediate solvate phases during crystallization. (e) Reducing the strength by an additional counter ion in the EDL (electrical double layer) leads to a faster collapse of the solvated system and turns directly into the desired structure. Reproduced from Ref. [56].

most effective strategies for managing the morphology of perovskite films. Showing the fundamental effect of A-site cation composition on perovskite precursor alignment and subsequent film formation, Flatken *et al.* elucidate the mechanism for controlling the colloidal stability of perovskite precursors and thus film morphology using alkali metals, as shown in Figure 2.11 [56].

2.2.2.2 *Crystal orientation of thin films*

Polycrystalline thin films are rich in properties, such as grain size distribution, grain shape, and crystal orientation, all of which influence the microstructure of perovskites to varying degrees. Grain crystal orientation may be as important as grain size for a particular perovskite film [60]. In $MAPbI_3$ films, a preferred (110) grain orientation (for the room-temperature tetragonal phase) is often associated with enhanced perovskite film performance. In general, for devices based on 3D structured perovskite thin films, performance and film properties are largely governed by the quality inherent in the grain orientations, rather than how these orientations interact with other device layers. However, the effect of crystal orientation is more pronounced in layered perovskite (e.g., 2D or Ruddlesden-Popper) films. For 2D perovskites, the morphology and crystal orientation of polycrystalline films directly affect the device performance, so understanding and control of the microstructure of perovskite films is crucial. Maria A. Loi *et al.* used a variety of techniques to reveal the microstructure of layered perovskite $(PEA)_2PbI_4$ blade-coated films, and found that the orientation of the grains made the [001] plane almost completely parallel to the substrate surface, while the [100] and [010] grain directions do not show any preferred alignment, which leads to excellent device performance [61].

2.2.2.3 *Grain boundaries of thin films*

The behavior of grain boundaries in polycrystalline halide perovskite solar cells is still poorly understood, the grain boundaries themselves are very rich and complex, and there is no general consensus on the overall effect of grain boundaries. Grain boundaries play an important role in most polycrystalline solar cells. In perovskite solar cells, the effect of grain boundaries on the open circuit voltage is not at all significant; most studies have shown that the hysteresis is mainly caused by defects at the grain boundaries; the grain boundaries definitely contribute to increasing the degradation rate of perovskite [62]. Vaynzof *et al.* found that small hidden grains accompanied by a large number of grain boundaries act as non-radiative recombination hotspots, suggesting that eliminating small grains can improve perovskite solar cells performance [63]. Park *et al.* employed first-principles calculations to show that enhanced structural relaxation of grain boundary defects leads to increased stability and deeper trap states. And show how partial passivation of the grain boundaries by halide mixing or extrinsic doping suppresses the formation of trap states close to the grain boundaries [64]. Mehdizadeh-Rad *et al.* have studied in detail the effect of grain boundary diameter on the operating temperature of perovskite solar cells. It was found that by increasing the grain boundary diameter, the operating temperature of the perovskite solar cells decreased at higher voltages and the open circuit voltage increased slightly [65].

2.2.3 *Vapor-based Deposition Techniques*

2.2.3.1 *Hybrid chemical vapor deposition*

Hybrid chemical vapor deposition affords advantages such as uniform deposition across large areas, being solvent-free, and bearing readiness for integration with other thin-film solar technologies to form tandem solar cells. For instance, Qiu *et al.* develop a fully vapor-based scalable hybrid chemical vapor deposition process for depositing Cs-formamidinium (FA)-mixed cation perovskite films (shown in Figure 2.12(a)), which alleviates the problem encountered when using conventional solution coating of main methylammonium lead iodide (MAPbI$_3$). And a large area solar module with efficiency approaching 10% and with a designated area of 91.8 cm^2 has been made [66]. On this basis, they further demonstrate a rapid HCVD (RHCVD) fabrication process for PSCs using a rapid thermal process, which not only significantly reduces the deposition time to less than 10 min, but also effectively suppresses hysteresis [67].

2.2.3.2 *Vapor deposition by dual-source*

Vapor deposition by dual-source uses two heating sources to heat metal halide and organic ammonium salt, respectively, under vacuum and involves deposition on the substrate at the same time. Further annealing the obtained perovskite film can increase the grain size to several hundred nanometers with good uniformity and flatness. For example, Liu *et al.* first reported the process of depositing perovskite thin films by the

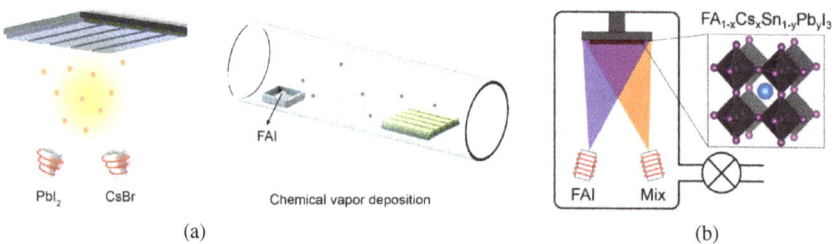

Figure 2.12.   (a) Schematic drawing showing the HCVD process of Cs$_{0.1}$FA$_{0.9}$PbI$_{2.9}$Br$_{0.1}$. Reproduced from Ref. [66]. (b) Schematic diagram of FA$_{1-x}$Cs$_x$Sn$_{1-y}$Pb$_y$I$_3$ film preparation process. Reproduced from Ref. [70].

dual-source vacuum evaporation method, and the efficiency of the pre-pared device reached 15% [68]. Similarly, Ball *et al.* used mixtures formed by melting metal halides as a single-crucible source of Cs, Pb, and Sn cations. Surprisingly, when this melt was co-evaporated with formami-dinium iodide (FAI), uniform and dense perovskite films in the family $FA_{1-x}Cs_xSn_{1-y}Pb_yI_3$ were formed. Inclusion of SnF_2 in the melt helped to regulate the perovskite's optoelectronic quality, leading to a steady-state power conversion efficiency of ~10% in a solar cell. The preparation pro-cess is illustrated in Figure 2.12(b) [69].

2.2.3.3 *Vacuum sequential deposition*

Vacuum sequential deposition (VSD) was developed because of the dif-ficulty in monitoring the organic ammonium salt deposition rate in the co-deposition process. The photovoltaic performance of devices pre-pared by VSD was found to depend significantly on the substrate tem-perature, in which metal halide was first deposited by thermal evaporation followed by vapor deposition of organic ammonium salt [72]. For example, devices based on $MAPbI_3$ perovskite thin films pre-pared by the sequential vapor-deposition method exhibited a PCE of 15.4% [73]. Li *et al.* used VSD combined with surface fumigation to successfully fabricate curved perovskite cells with device efficiencies as high as 19.56% [74].

2.2.4 *Solution Chemistry Deposition of Perovskite Films*

2.2.4.1 *One-step solution-processing method*

Among the methods for preparing perovskite thin films, the one-step method is the simplest by comparison. Taking $MAPbI_3$ as an example, PbI_2 and MAI are mixed and dissolved in a suitable solvent (such as GBL, DMF) to form a precursor solution, and then deposited by spin coating to form thin films. Kojima *et al.* first prepared perovskite thin films by one-step spin coating (shown in Figure 2.13), but the film-forming properties were poor [75]. Improvements were made by changing the ratio of raw materials, adding additives, changing solvents, introducing anti-solvents, and changing annealing temperatures.

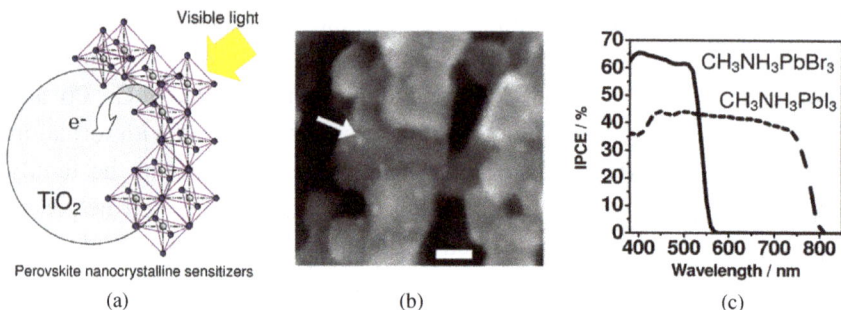

Figure 2.13. (a) Crystal structures of $CH_3NH_3PbX_3$ as visible-light sensitizer compounds. (b) SEM image of particles of nanocrystalline $CH_3NH_3PbBr_3$ deposited on the TiO_2 surface using one-step method. The arrow indicates a particle, and the scale bar shows 10 nm. Reproduced from Ref. [75] .

As an example of changing the ratio of raw materials to improve the film properties, Xu *et al.* introduced dry HCl gas into the precursor solution as Cl⁻ addition source. As the molar equivalent of Cl⁻ to Pb^{2+} in the precursor solution increases from 0.0 to 3.0, the perovskite film obtained by spin coating will become denser, the film will become smoother, and the grain size in the film will grow [76]. Similarly, Tang *et al.* used lead acetate as an additive to the perovskite precursor solution to regulate the preparation process of perovskite films to obtain high-performance perovskite solar cells [77]. The addition of lead acetate significantly delayed the crystallization process of perovskite films. In the crystallization process, lead acetate and amino group formed a stable intermediate state through hydrogen bond, which greatly delayed the crystallization process of perovskite, the grain size increased from 200 nm to 500 nm, and the defect density decreased by 10 times [78]. Zhao *et al.* developed a method to enhance device performance by adding trace amounts of CH_3NH_3Br (methylamine bromide) to lead acetate and CH_3NH_3I (methylamine iodide) precursor fluids [79]. Through optimization and characterization, the appropriate addition of a trace amount of CH_3NH_3Br can not only effectively improve the surface morphology and crystallinity of the perovskite film, but also adjust the optical and electrical properties of the film.

Additives can also have a great influence on the film formation process. Wu *et al.* introduced two additives, the ionic liquid additive methylammonium acetate (MAAc) and the molecular additive thiosemi-carbazide (TSC). By introducing 10–15 mol% MAAc, very uniform MAPbI$_3$ perovskite films can be generated by a simple one-step spin coating method. The addition of a small amount of TSC (3–5 mol.%) can effectively increase the crystal size of the perovskite [80]. Miao *et al.* prepared three-dimensional CsPbI$_3$ thin films with cuboid microcrystalline structure *in situ* by a one-step method. By introducing bisamine cations, propylenediamine hydroiodide (PDAI), into the perovskite precursor solution, the crystallization rate of perovskite is slowed down significantly, resulting in improved grain size and crystallization. PDAI can also form a passivation layer on the surface of the perovskite to further reduce the defect density and improve the fluorescence quantum yield and device efficiency of the film [81]. Likewise, Park *et al.* found that the isopropanol solvent added in the one-step method would chemically react with CH$_3$NH$_3$Cl, and the generated isopropylammonium cation could stabilize the α-FAPbI$_3$ phase. A certified efficiency of 23.9% for PSCs prepared by a one-step method is achieved by adding IPAm HCl to the perovskite precursor solution [82].

Common solvents for perovskite precursor solutions are dimethylformamide (DMF), G-butyrolactone (GBL), dimethyl sulfoxide (DMSO), and N, N-dimethylacetamide (DMA). They can be removed by thermal annealing or drying in air and inert atmosphere, causing supersaturation of the perovskite material during the solvent removal process, thus leading to its nucleation and crystal growth. This one-step solution processing method is simple and easy to implement [83]. The interaction between perovskite materials and solvents often has a great influence on the nucleation and growth of crystals. For example, Shan *et al.* compared the use of N, N-dimethylacetamide (DMAc) and DMF as solvents to prepare perovskite films [84]. In the transformation process of perovskite films, they found that DMF can enter the layered crystal structure of lead iodide together with methylamine iodide, adjusting the growth rate of the film to make grain size more uniform. In this way, they obtained planar-structured devices with high fill factor. Similarly, Seok *et al.* mixed DMSO with GBL as a solvent to generate a solid intermediate phase during the spin coating process, so that the crystal growth does not occur with strong aggregation and high-quality thin films can be obtained [85]. In addition, Chao *et al.*

used an ionic liquid solvent methylamine acetate (MAAc) to boost the photoelectric conversion efficiency (PCE) of perovskite cells to >20% by a simple one-step method in humid air without any additives [86].

In addition, there are other related preparation methods. For example, Zhang *et al.* developed a room temperature natural drying film formation process, which can prepare high-quality MAPbI$_3$ films without anti-solvent and the above-mentioned auxiliary crystallization methods. A low-boiling methylamine/ethanol/acetonitrile solvent system is used to add a drop of precursor solution in the center of the substrate. The solution can spread spontaneously at room temperature, and the MAPbI$_3$ film can be obtained by natural drying [87].

2.2.4.2 *Two-step solution-processing method*

The conventional two-step method is to first prepare a metal halide film by spin coating and then let it react with an organic ammonium salt or cesium halide to obtain the perovskite film. Based on the different preparation processes of the second step, the method can be subdivided into two-part immersion method, two-part spin coating method, and vapor-assisted solution method.

The two-step immersion method is to soak the metal halide film in the solution of organic ammonium salt to obtain the perovskite film. For example, Graetzel *et al.* reported the preparation of CH$_3$NH$_3$PbI$_3$ perovskite thin films by a two-step solution immersion method for the first time [88]. First, a PbI$_2$ solution was spin-coated on a mesoporous TiO$_2$ film, and a PbI$_2$ film was obtained after heat treatment. Then, the PbI$_2$ film was soaked in a certain concentration of CH$_3$NH$_3$I isopropanol. In the solution, CH$_3$NH$_3$I and PbI$_2$ chemically react to form dark brown perovskite, which can be obtained by heating and annealing. The PCE of the final assembled perovskite battery device is 15%. Similarly, Zheng *et al.* adopted precise wetting steps and higher immersion reaction temperature to accelerate the growth of perovskite crystals, making the interface between perovskite and hole transport layer rough and continuous [89]. The light scattering generated by this rough interface enhances the absorption of the device in the long-wavelength direction, and is also conducive to the transfer of charges, improving the short circuit current and PCE of the device.

The two-step spin coating method is to first drop an organic amine salt (or cesium halide) during the high-speed rotation of the precursor film and then thermally anneal it to form perovskite. Compared with the two-step dipping method, the two-step spin coating method avoids the long-term immersion that may damage the perovskite. For instance, as demonstrated by Xiao *et al.*, after the isopropanol solution of the organic amine salt drops on the precursor film, the subsequent thermal annealing makes the organic amine salt remaining on the surface of the precursor film enter the precursor film through free diffusion and fully react with PbI_2. The iodide ions in the organic amine salt are inserted between the PbI_2 layers to form PbI_3^-, and then react with the organic amine salt ions to form uniform and dense perovskite films [90]. Park *et al.* compared the two-step spin coating in mesoporous devices and studied the difference between the film morphology obtained by the coating method and the one-step spin coating method [91]. The film prepared by the one-step spin coating method is prone to agglomeration, making the film very uneven, and the mesoporous TiO_2 in some areas is not filled. With the two-step spin coating method, the thin films prepared are very uniform and the devices have good reproducibility.

The vapor-assisted solution method combines solution spin coating and vapor deposition. For example, PbX_2 was first deposited on the substrate by solution method and CH_3NH_3X solid was then heated and evaporated, with the vapor diffused and penetrated slowly and combined with PbX_2 to form a perovskite film, as shown in Figure 2.14 [92]. Different from other traditional solution processing methods, the vapor-assisted solution processing method avoids the dissolution and solvation of the film during the growth process, inhibits the formation of crystal nuclei, and enables the film to reorganize rapidly to obtain dense high-quality perovskite films.

The energy conversion efficiency of $MAPbI_3$ solar cells prepared by vapor-assisted solution method for the first time is 12.1% [93]. Since the steam-assisted solution process preparation method was reported, low-pressure vapor chemical vapor deposition and hybrid chemical vapor deposition methods have been intensively studied, respectively [94,95]. For example, *in situ* tubular chemical vapor deposition (ITCVD) was used to fabricate large-area perovskite films with good optical quality for solar cells, obtaining 12.2% conversion efficiency [96].

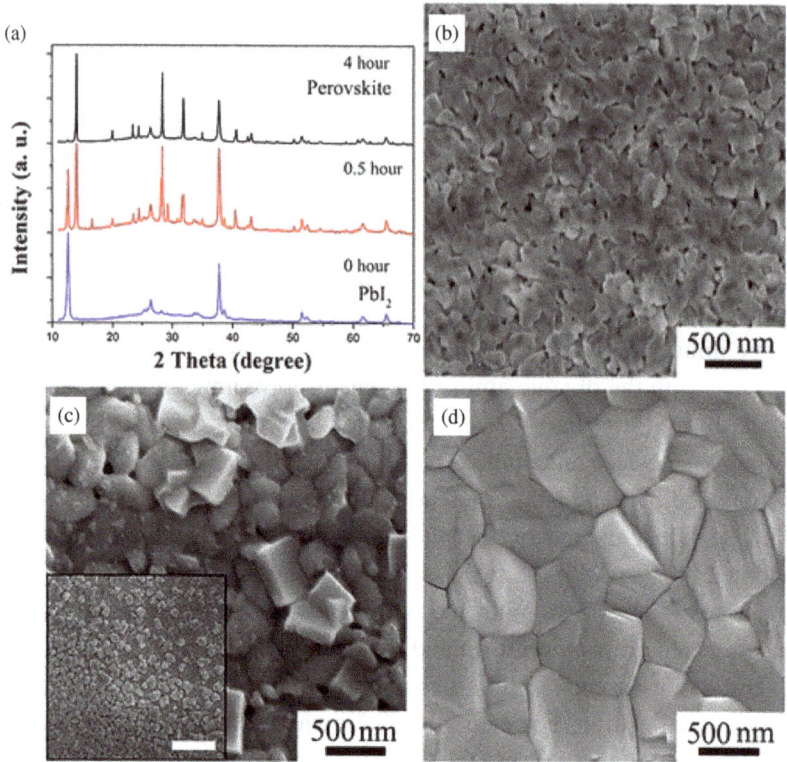

Figure 2.14. Time evolution characterization of the perovskite thin film by annealing ~200 nm PbI_2 film in the presence of CH_3NH_3I at 150°C in N_2 atmosphere using the vapor-assisted solution method: (a) XRD patterns of the film annealed at 0, 0.5, and 4 h; (b–d) top-view SEM images of (b) the initial stage at 0 h, (c) the intermediate stage at 0.5 h (inset: wider view, scale bar 3 μm), and (d) the post stage at 4 h. Reproduced from Ref. [92].

In a related study, Li *et al.* prepared 2D/3D organometallic halide perovskites for the first time by low-pressure vapor-assisted solution processing [97]. The specific method is to react phenylethylammonium iodide (PEAI)-doped lead iodide (PbI_2) substrate with methylammonium iodide (MAI) vapor placed in a low-pressure heating furnace. Scanning electron microscopy images and Kelvin probe force microscopy maps indicate that PEAI-containing 2D perovskite large grains surround the 3D $MAPbI_3$ perovskite grains, which is favorable for $MAPbI_3$ grain growth. The doping ratio of PEAI has an important influence on the crystal structure, surface morphology, grain size, and device performance. With PEAI/PbI$_2$ = 0.05, the PCE of the 2D/3D organometal halide perovskite device is 19.10%, with a fill factor of 80.36% and good device stability.

2.2.4.3 *Multistep solution-processing method*

A multistep solution-processing method is mostly used in the preparation of CsPbBr$_3$ films. Due to the low solubility of CsBr in precursor solutions, solution-processed CsPbBr$_3$ films are typically discontinuous and porous, hindering the performance of the resulting perovskite film. To address this issue, Tang *et al.* developed a multi-step solution processing method, in which they fabricated high-purity inorganic CsPbBr$_3$ perovskite films [98]. The phase transitions of CsPb$_2$Br$_5$ to CsPbBr$_3$ and Cs$_4$PbBr$_6$ were gradually changed by varying the number of deposition cycles of the CsBr solution to achieve vertical and monolayer aligned grains. And, a proto-type cesium lead bromide (CsPbBr$_3$) perovskite solar cell with a fluorine-doped tin oxide (FTO)/CsPbBr$_3$/carbon structure was constructed by a multi-step solution-process deposition technique with an efficiency as high as 4.1% [99]. Similarly, a facile and modified sequential spin-coating method is introduced by Liu *et al.* to prepare high-crystallinity, pinhole-free CsPb$_2$Br$_5$-CsPbBr$_3$ perovskite films with an average grain size of ≈1 μm [100]. For the solar cells, the film morphology is key to their per-formance. Especially, pinholes generated during spin coating adversely affect the solar cell performance. Achoi *et al.* have demonstrated a method to prepare pinhole-free methylammonium bismuth iodide (MBI) film by multi-step spin coating [101].

2.2.4.4 *Other methods*

In addition to the above preparation methods, other techniques such as doctor blading, slot casting, spray coating, and inkjet printing have also been reported. Doctor blading is a deposition method relying on a similar principle to that of spin-coating. By forming a thin layer of solution from which the solvent evaporates uniformly, the precursors reach the satura-tion point and precipitate in an even manner as the film dries. Unlike spin-coating, however, the excess solution is removed by sweeping a blade across the substrate at a precisely fixed height rather than by spinning rapidly. This method is, therefore, ideally suitable for roll-to-roll manufacturing, as the solution can first be deposited on a moving substrate and then swept under a stationary blade [59]. Deng *et al.* used this method to prepare perovskite thin films, and when the effective area of the battery was 7.25 mm^2, the efficiency of the battery could reach 15.1% [102]. Slot coating has the advantages of low cost, high throughput, and good continuity. It can precisely control the liquid flow and coating speed. It is currently one of the most promising deposition techniques. However, the challenge with this technique is how to control the uniformity of the large-area film and reduce the defect density of the film. In addition, improving the uniformity and reducing defects on the film surface are the keys to the successful fabrication of high-efficiency large-area perovskite devices.

In response to the above problems, Du *et al.* developed high-pressure nitrogen extraction and ionic liquid interface passivation technology combined with slot coating deposition method to prepare large-area battery components [103]. For spray coating, a perovskite precursor solution is ejected through a nozzle, aerosolizing it. The resulting droplets travel to the substrate, where the solvent evaporates, depositing the perovskite precursors, which may then be annealed to generate the final film. Huang *et al.* fabricated a perovskite cell with an effective area of 1 cm^2 by ultrasonic spraying, and the efficiency could reach 13.09%, proving the superiority of the perovskite cell prepared by the spraying method [104]. Like inkjet printing on paper, inkjet printing technology has the advantages of being able to produce under normal pressure, less ink dripping onto the outside of the substrate, and improving material usage efficiency. For instance, Wei *et al.* fabricated a planar-structured perovskite cell with a structure of $TiO_2/CH_3NH_3PbI_3/C$ by an inkjet printing method, with an efficiency of 11.6% [105].

2.3 Synthesis of Metal Halide Perovskite Nanocrystals

2.3.1 *Background*

The first regular wet chemical synthesis was reported by Perez–Prieto and co-workers who produced 6 nm MAPbBr$_3$ NCs in 2014 [106–111]. With the assistance of ammonium ions, those NCs could be stabilized in non-polar organic solvents. This was followed by two reports by Zhong [112] and Kovalenko [113] groups who introduced the LARP method for hybrid perovskite NCs and the hot-injection method for all-inorganic perovskite NCs, respectively. We will focus on the two most developed liquid-phase methods: the ligand-assisted reprecipitation (LARP) method and the hot injection (HI) method for the synthesis of colloidal MHP NCs. The LARP method can be employed as a more cost-effective alternative, as it can generate high quality perovskite NCs in an ambient atmosphere at room temperature. The HI route requires high temperatures and an inert atmosphere, which inevitably increases the cost and could limit the output in mass production. One significant advantage of the reprecipitation method over HI method is the relative simplicity, typically requiring only simple air-free conditions. As such, it is more easily scalable and implementable for the industrial scale preparation of perovskites.

2.3.2 *Growth Mechanism*

In growing MHP NCs, critical factors include nucleation, growth, and ligand passivation to generate defect-free and size-controlled products. Understanding the role of these factors in the formation of MHP NCs is crucial to their device applications. In the LARP method, polar solvents such as DMF or DMSO, acting as a "good" solvent, are used to dissolve all of the precursor salts and coordinating ligands such as oleic acid (OA), oleylamine (OLA), and octylamine (OAm), and then they are injected into a non-polar "bad" solvent such as toluene or hexane [114]. The perovskite NCs will be formed right after injection, while bigger microcrystals may also be formed alongside them, depending on the conditions. Due to the low formation energy of the organic–inorganic perovskites, this synthesis can be conveniently conducted at room temperature. The synthesis of all-inorganic perovskite NCs (with higher stability) required the HI method [113]. All-inorganic $CsPbX_3$ NCs have higher formation energies, so they benefit from preparation at higher temperatures, mostly at 120–200°C. They can be formed by swiftly injecting Cs-oleate (obtained through the reaction between Cs_2CO_3 and OA) into a non-coordinating solvent (such as 1-octadecene, ODE) containing PbX_2 and ligands under N_2 atmosphere.

Rapid nucleation occurs immediately upon the injection of the required ingredients producing small nuclei. The size of the NCs may be reduced by decreasing the reaction temperature.

2.3.3 *Ligand-assisted Reprecipitation (LARP)*

LARP involves the precipitation of perovskite NCs at room temperature directly from the addition of precursors and ligands dissolved in polar solvents such as DMF or dimethyl sulphoxide (DMSO), or γ-butyrolactone (GBL) into a non-polar solvent such as toluene. The crystallization process in LARP is caused by the supersaturation induced by the solubility change with solvent mixing. The first report on LARP synthesis of hybrid organic–inorganic LHP was by Papavassiliou *et al.* in 2012 [115]. They solubilized $MAPbX_3$ (X = Br, Cl or I) salts in DMF (or acetonitrile) and dropped the corresponding solutions in toluene (or in a mixture of toluene and polymethyl methacrylate) to obtain nanocrystalline/microcrystalline $MAPbX_3$.

2.3.3.1 *$MAPbX_3$ nanocrystals*

The LARP approach was demonstrated and systematically studied in 2015 by Zhong *et al.* for synthesizing $MAPbBr_3$ NCs, and a schematic illustration of the reaction system and process is given in Figure 2.15 [112]. The basic procedure includes two main steps. First, a precursor solution is prepared by dissolving $MA^+/Cs^+/FA^+$, Pb^{2+}, and halogen ion precursors and capping ligands (alkylamine and carboxylic acid) in polar solvents such as DMF or DMSO. Second, the precursor solution is injected into anti-solvents such as acetone/toluene/hexane at room temperature under vigorous stirring. By replacing the bromide anion with chloride or iodide, or by using mixtures of the three anions, the authors were able to obtain perovskite NC dispersions emitting throughout the full range. In order to understand the specific role of alkyl amines and carboxylic acids, different amines (hexylamine, OAm, dodecylamine, hexadecylamine) and acids (OA, octanoic acid, or butyric acid) were examined. Interestingly, the formation of $MAPbBr_3$ NCs could be achieved even without the use of amines, but there was no control over the size of the crystals. On the other hand, the exclusion of carboxylic acids from the synthesis resulted in

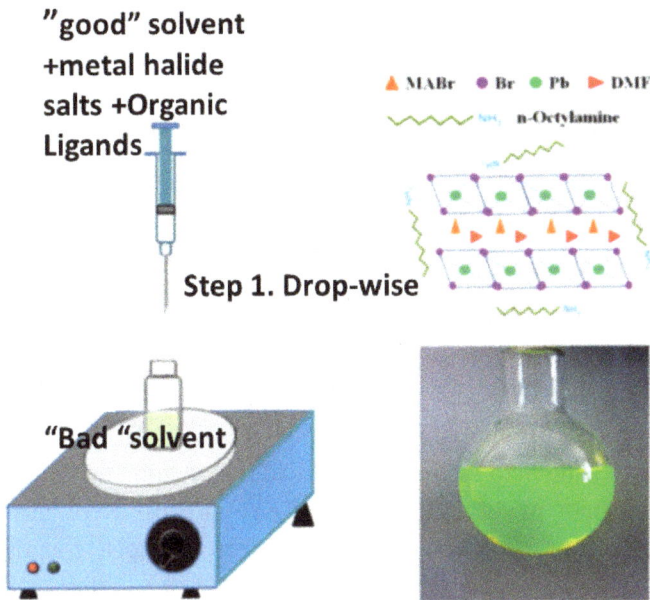

Figure 2.15. Schematic illustration of the reaction system and process for LARP technique. Reproduced from Ref. [112].

aggregation of NCs. It was concluded that the amines play a key role in controlling the kinetics of crystallization and mainly contribute to the size control of NCs formed. Whereas the organic acids suppress aggregation effects and ensure their colloidal stability. Later, this same approach was exploited and further optimized by different groups [116–118]. For example, Huang *et al.* demonstrated that the size and size distribution of MAPbBr$_3$ NCs could be better controlled by introducing OLA together with OA and varying the reaction temperature [116]. Arunkumar *et al.* demonstrated that it is possible to dope MAPbX$_3$ NCs with Mn^{2+} simply by adding MnCl$_2$ to the precursor solution [117]. Other ligands were subsequently proved to work efficiently in the LARP synthesis of MAPbBr$_3$ NCs. Luo *et al.* produced size tunable MAPbX$_3$ (X = Cl, Br, or I) NCs by employing two branched capping ligands, (3-aminopropyl) triethoxysilane (APTES) and polyhedral oligomeric silsesquioxane (POSS) PSS-[3-(2-aminoethyl)amino]- propylheptaisobutyl-substituted (NH$_2$-POSS) [119]. It was proposed that, compared to straight-chain ligands, APTES and NH_2-POSS offer greater control over particle size and uniformity, which is attributed to their strong steric hindrance and can protect PNCs from being redissolved by DMF. Veldhuis *et al.* introduced the use of benzoyl alcohol as an auxiliary ligand together with OAm and OA, which was found to accelerate the reaction kinetics and to improve the optical properties of the resulting NCs [120].

Later on, Luo *et al.* employed peptide-like molecules,12-aminodode-canoic acid, as bidentate ligand in the synthesis of MAPbBr$_3$ NCs. Peptides with both $-NH_2$ and $-COOH$ groups enabled a good control over the size of the resulting NCs [122]. Similarly, Wang *et al.* demonstrated a n L-cysteine, consisting of $-NH_2$ and $-COOH$ and S^{2-} groups-assisted LARP approach for preparing monodisperse $CH_3NH_3PbBr_3$ PNCs and then assembled them to fabricate stable and highly luminescent supercrystals [123]. Minor modifications to the LARP process were also demonstrated to further optimize the process. For example, Shamsi *et al.* designed an alternative LARP approach in which PbX$_2$ salts are dissolved in NMF, rather than in the typical DMF, together with OLA and OA. The solution is heated up to 100°C for 10 min and finally added dropwise at room temperature to a poor solvent (such as dichlorobenzene or chloroform) [29]. The advantage of this approach is that MA$^+$ ions are formed *in situ* during the heating step (by a transamination reaction), therefore there is no need to first synthesize MAX salts. The same approach was also used to yield bulk crystals at room temperature with no need for an anti-solvent. Dai *et al.* mixed both good solvents and poor solvents using a spray, producing MAPbBr$_3$ NCs with a good size distribution [124]. In this case, the precursor solution (MABr, PbBr$_2$, OA, OAm in DMF) was sprayed onto a poor solvent (toluene). The micrometer-sized droplets of the sprayed solution provide a large contact surface area between the two solutions, allowing for a more homogeneous mixing.

Soon after the development of the LARP synthesis of MAPbX$_3$ NCs, efforts were made to optimize the PL of the materials and to control their size and shape. In 2015, Sichert *et al.* reported a LARP synthesis of MAPbBr$_3$ nanoplates (NPLs) in which no carboxylic acid was employed, with only octylammonium and MA$^+$ ions used as ligands. The systematic increase in the octylammonium/MA$^+$ ratio caused a reduction in the thickness of the resulting NPLs. The exception to this was the use of octyl-lammonium only, when "single-layered" NPLs were observed, similar to what was reported in the studies on layered perovskite microcrystals [125]. Analogous results were achieved in 2016 by different groups. Kumar *et al.* demonstrated control over the thickness of $MAPbBr_3$ NPLs down to one monolayer using OA and OAm [126]. Cho *et al.* carried out a systematic study on the LARP synthesis of $MAPbBr_3$ NPLs in which OA and several alkylamines with different chain lengths (such as butyl-amine, hexylamine, OAm, dodecylamine, and OLA) were used [127]. Both the chain length and the concentration of alkylamines played an essential role in determining the thickness of $MAPbBr_3$ NPLs. A high concentration of alkylammonium cations can efficiently passivate the surface of $MAPbBr_3$ NCs, preventing their growth along the vertical direction and yielding NPLs with tunable thicknesses. In 2017, Levchuk *et al.*, using the mixture of OLA and OA, achieved a good control over the thickness of $MAPbI_3$ NPLs by introducing chloroform as the bad solvent [128]. Ahmed *et al.* proposed using pyridine, together with OLA and OA, as an effective co-ligand for finely tuning the thickness of $MAPbBr_3$ NPLs [121]. DFT calculations revealed that pyridine molecules can bind to the Pb^{2+} ions that are present on the surface of the growing nanostructures, forming dative $N \rightarrow Pb$ bonds and slowing down the vertical growth rate, thus leading to the formation of 2D nanostructures (see Figure 2.16).

Huang *et al.* conducted a systematic study on how the amount of ligands (OLA and OA), the precursors/ligands ratio, and the reaction temperature affect $MAPbBr_3$ NCs [129]. Similar to what was reported by Cho *et al.*, they found that a fine-changing of these parameters resulted in a control over the size, and, thus, of the quantum confinement of $MAPbBr_3$ NCs [127]. Working with a high ligand/precursor ratio, they observed the formation of small NCs, while a high precursor/ligand ratio produced polydisperse micro- and NCs. At intermediate ligand/precursor ratios, the nucleation and growth of the NCs could be better controlled: the synthesis under such conditions produced small NCs that grew over time. In this way, alkylammonium ligands (or pyridine) can significantly passivate the surface of $MAPbBr_3$ NCs, preferentially slowing down the vertical growth rate and allowing NCs to grow only along the lateral directions.

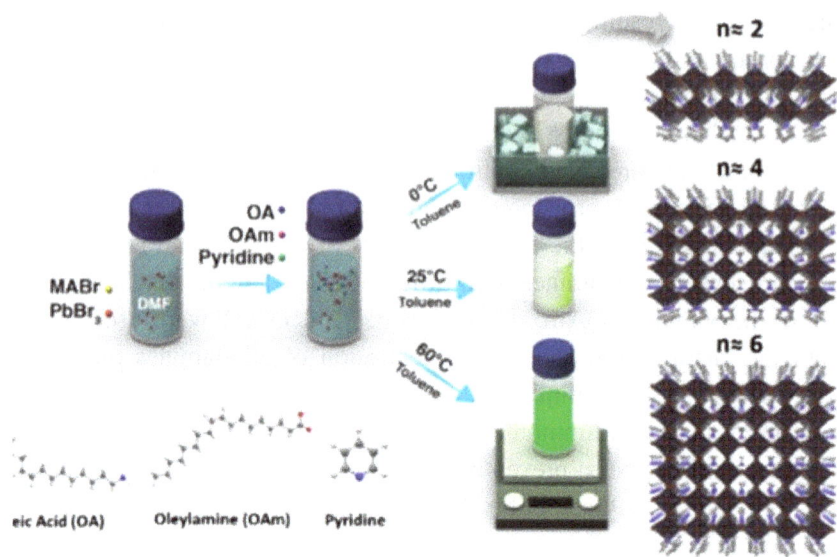

Figure 2.16.   Schematic illustration of a LARP synthesis of MAPbBr$_3$ nanostructures using oleic acid (OA), oleylamine (OLA), and pyridine as ligands. Reproduced from Ref. [121].

2.3.3.2 *FAPbX$_3$ nanocrystals*

In 2016, Weidman *et al.* reported a LARP synthesis of both all-inorganic and organic–inorganic ABX$_3$ (A = Cs, MA, or FA; B = Pb or Sn) NPLs with 1 or 2 monolayer thicknesses [130]. The precursors, AX and BX$_2$ salts, were dissolved in DMF together with octylammonium and butylam-monium halides (used in a 1:1 ratio). The precursor solution was added dropwise to toluene under vigorous stirring at room temperature to trigger a formation of NCs. To achieve colloidal stability and thickness homoge-neity, an excess of ligands was used (in a 10:2:1 ratio of ligands/BX$_2$/AX). Later on, Perumal *et al.* reported the synthesis of FAPbBr$_3$ NCs, in which FABr and PbBr$_2$ were dissolved in DMF and added dropwise to a solution consisting of toluene, butanol, OAm, and OA with continual stirring [131]. The first LARP synthesis of FAPbX$_3$ (X = Cl, Br, or I) NC systems was reported by Levchuk *et al.* [114]. The synthesis relies on the rapid injection of a precursor solution, which is prepared by dissolving PbX$_2$ and FAX (X = Cl, Br, or I) salts in DMF, OA and OLA into chloroform at room temperature. With dependence on the OLA/OA ratio, either nanocubes or NPLs with controlled thickness (ranging from 2 to 4 mon-olayers) were produced (Figure 2.17). Toluene prevented the formation of FAPbI$_3$ NCs, and led to the immediate agglomeration of FAPbBr$_3$ or FAPbCl$_3$ NCs. Minh *et al.* reported the synthesis of FAPbX$_3$ nanocubes, introducing PbX$_2$–DMSO complexes as novel precursors [132]. All the precursors (the PbX$_2$–DMSO complex and FAX) were dissolved in DMF together with OLA, and the corresponding solution was added to a mix-ture of toluene and OA. The formation of FAPbX$_3$ NCs is believed to occur via an intramolecular exchange reaction in which alkylammonium halides can replace DMSO molecules in the starting PbX$_2$–DMSO com-plex. Kumar *et al.* reported a slightly modified LARP procedure to create FAPbBr$_3$ NPLs with a very high PLQY, which could be used in LED devices [133]. The difference here is that FABr and PbBr$_2$ salts were dis-solved in ethanol and DMF, respectively, to form two different polar

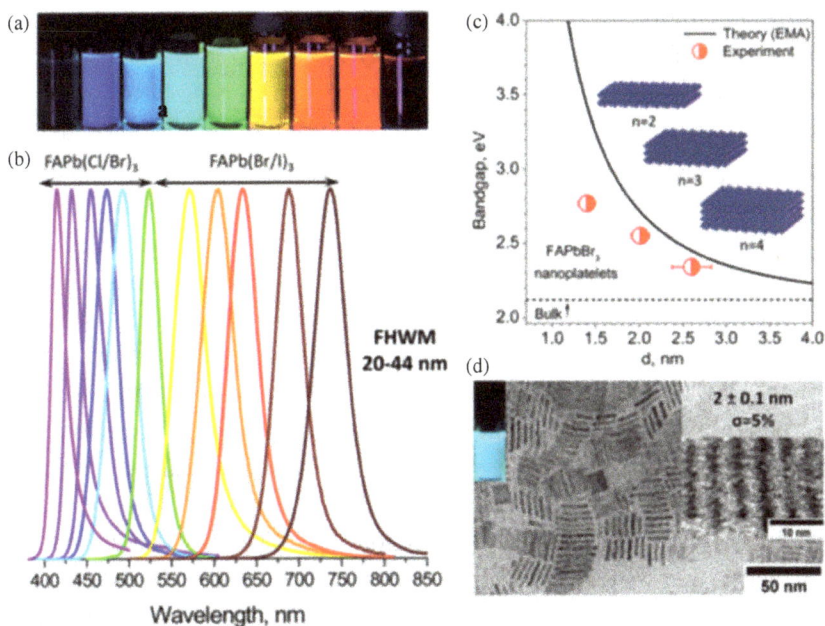

Figure 2.17. (a) Photograph of FAPbX$_3$ NC dispersions under UV-light. (b) The corre-sponding PL emission curves. (c) Theoretical effective mass approximation and experimen-tal band gaps of FAPbBr$_3$ NPLs as a function of their thickness and (d) TEM characterization of vertically stacked FAPbBr$_3$ NPLs. All panels are reproduced from Ref. [114].

solutions, which were then added simultaneously to a mixture of toluene, OA, and OAm [133].

2.3.3.3 *All-inorganic CsPbX$_3$ nanocrystals*

In 2016, Li *et al.* showed that all-inorganic perovskite colloidal NC systems could also be produced using the LARP approach at room temperature. Their synthesis is similar to the one used for the organic–inorganic MAPbX$_3$ NC systems, with the only difference being that CsX is used rather than MAX in the precursor solution [134,135]. Moreover, the inorganic salts (CsX and PbX$_2$) are dissolved in DMF together with OLA and OA, which are then added to toluene, resulting in an immediate formation of NCs. Later, Seth *et al.* extended this approach to achieve control over the morphology of CsPbX$_3$ NCs [136]. The shape of the NCs could be controlled by varying the bad solvent (from toluene to ethyl acetate), the relative amount of ligands, and the reaction time: the use of ethyl acetate promotes the formation of quasi-cubic QDs, NPLs, or nanobars, while toluene can be used for the preparation of nanocubes, NRs, or NWs (Figure 2.18) [136]. These different morphologies are attributed to the different ligand/non-polar solvent interactions. Since ethyl acetate is more polar than toluene, it can act both as a solvent and a nucleophile, causing some OLA molecules to leave off the surface of the growing nuclei, thus undergoing an oriented attachment. In toluene, if the concentration of OLA is high enough, the NC surfaces become more protected in all directions, which consequently prevents both their attachment/merging and growth. However, if only a small amount of OLA is used in toluene, the NCs can grow anisotropically at longer reaction times, in the form of NRs and NWs, most likely as a consequence of an incomplete surface passivation of the facets. In 2017, Kostopoulou *et al.* reported a LARP synthesis of CsPbBr$_3$ NWs with micron-sized lengths [137]. The key feature of their approach is that they use anhydrous solvents and a low temperature with the precursor solution (CsBr, PbBr$_2$, DMF, OLA, and OA) dropped into anhydrous toluene, which is kept in 0°C. Immediately after the injection, the resulting product consists of small-length, bullet-like NRs, which evolve into NWs with a width of 2.6 nm after 24 h at RT, and thicker NWs

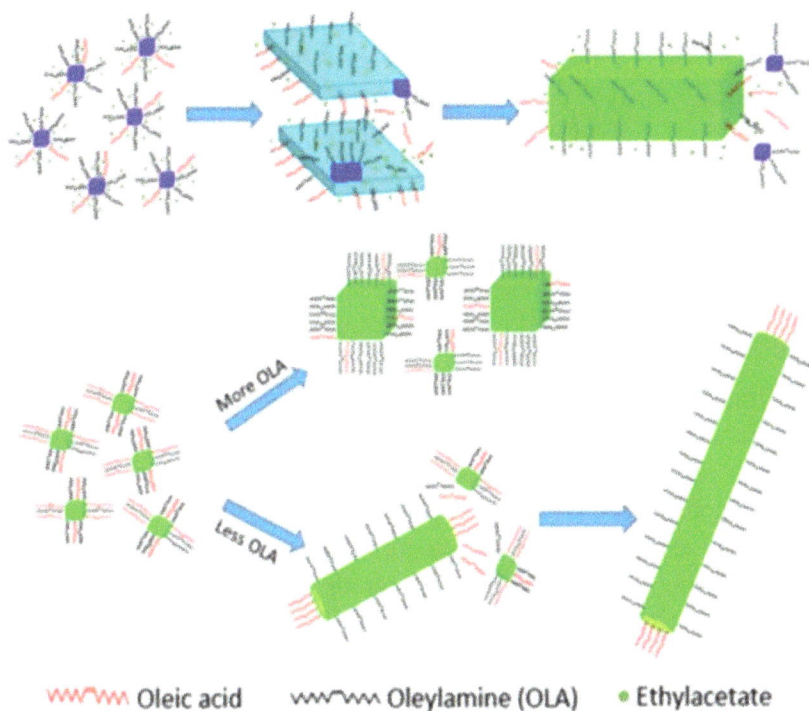

www Oleic acid   www Oleylamine (OLA)   • Ethylacetate

Figure 2.18. Sketch of the mechanism proposed for the formation of NPLs (cyan) and nanobars (green) in ethyl acetate and larger nanocubes, nanorods, and NWs from smaller nanocubes in toluene. Reproduced from Ref. [136].

with a diameter of 6.1 nm after 1 week. In 2018, Zhang *et al.* deliberately introduced water to the LARP synthesis of $CsPbBr_3$ NCs, demonstrating its influence on the growth rate and the shape of the resulting crystals [138]. Both H_3O^+ and OH^- are suggested to act as surface ligands with a higher activity than those of oleylammonium and oleate species, causing the perovskite NCs to grow in different directions.

2.3.3.4 *Mixed A-cation $APbX_3$ nanocrystals*

To tune the bandgap and optical properties of $APbX_3$ NCs, one can vary the halide composition or the A cation [139]. For instance, Mittal *et al.*, for example, successfully tuned the bandgap of $APbBr_3$ NCs from 2.38 to 2.94 eV by varying the composition of the A site from pure MA to pure ethylammonium (EA) [140]. Similar studies have been reported on $Cs_{1-x}FA_xPbX_3$ NCs [141,142], and mixed Cs and MA ions to form $MA_{1-x}Cs_xPbBr$ NCs using mixed Cs and MA cations [143]. Zhang *et al.* reported a LARP synthesis of mix-organic-cation $FA_xMA_{1-x}PbX_3$ NCs, achieving continuously tunable PL emission from 460 to 565 nm [144].

2.3.3.5 *CsPb$_2$Br$_5$ NCs*

In 2016, Wang *et al.* reported colloidal synthesis of perovskite-related $CsPb_2Br_5$ NPLs by adopting the LARP approach [145]. Two different precursor solutions were prepared in DMF, one containing $PbBr_2$ and hexylammonium bromide, and another containing CsBr. These solutions were added to toluene to trigger the formation of the final NCs. The synthetic procedure could not be used for the preparation of pure phase $CsPb_2Cl_5$ and $CsPb_2I_5$ NCs by substituting bromide salts ($PbBr_2$, CsBr, and HABr) with the corresponding chloride or iodide ones. Ruan *et al.* subsequently optimized the LARP strategy in order to control the morphology of $CsPb_2Br_5$ NCs [146]. This was achieved by introducing alkyl thiols (octanethiol) as ligands together with either alkylamines (OLA) or carboxylicacids. More precisely, under their experimental conditions, $CsPb_2Br_5$ NWs were obtained by employing octanethiol and OLA, while $CsPb_2Br_5$ NCs required the use of alkyl-thiols and carboxylic-acids. The following year, the same authors demonstrated that their approach could be used for direct synthesis of tetragonal $CsPb_2X_5$ (X = Cl, Br, or I) NWs, even in mixed halide compositions [147].

2.3.3.6 *Pb-free perovskite-related nanocrystals*

There are very few examples of Pb-free perovskite-related NCs produced by the LARP strategy. In 2016, Leng *et al.* proposed a LARP synthesis of $MA_3Bi_2X_9$ (X = Cl, Br, or I) perovskite-related NCs [148], and, $Cs_3Bi_2X_9$ (X = Cl, Br, or I) NCs [149]. At first, octane was employed as the bad solvent, while DMF and ethyl acetate were used as good solvents for

MAX and BiX salts, respectively [148]. Subsequently, they used DMSO to solubilize both CsX and BiX$_3$ salts and ethanol to trigger a recrystallization [149]. As in the case of Pb-based perovskite NCs, the amount of surfactants (OAm and OA) was observed to play a major role in controlling the LARP synthesis of Cs$_3$Bi$_2$X$_9$ NCs. Interestingly, the use of toluene as a bad solvent, instead of ethanol, produced unstable NC solutions. In 2017, Zhang *et al.* adopted the LARP method to prepare blue-emitting Cs$_3$Sb$_2$Br$_9$ NCs. They dissolved SbBr$_3$ and CsBr salts in either DMF or DMSO in the presence of OLA to form a clear precursor solution, which was then dropped into a mixture of octane and OA [150].

Although the LARP method enables a direct synthesis of many different perovskite systems at RT under air, the LARP method also has some weak points. Perovskite NCs are very sensitive to polar solvents, therefore the polar solvents that are normally used in the LARP synthesis, for example DMF, can easily degrade and even dissolve the CsPbX$_3$ NCs, especially the CsPbI$_3$ ones [151]. Precursor–polar solvent interactions were observed to play an important role in the formation of defective perovskite NCs. To elucidate this "solvent effect", Zhang *et al.* investigated the effects of using different polar solvents on the crystallization of MAPbI$_3$ NCs [152]. They found that PbI$_2$ generates stable intermediates in coordination solvents like DMSO, DMF, and tetrahydrofuran (THF), but it does not form complexes when it is dissolved in non-coordinating solvents, like GBL and MeCN. These different binding motifs have, in turn, different impacts on the crystal structure of the synthesized NCs, with the strong bonding between PbI$_2$ and the coordinating solvents leading to the formation of defective MAPbI$_3$ NCs that contain residual solvent molecules on the surface and iodine vacancies in the bulk, while non-coordinating solvents are unable to lead to crystallization of defect-free MAPbI$_3$ NCs. Furthermore, DMF and DMSO have a high boiling point and are toxic, therefore not suitable for large-scale production.

2.3.4 *Hot Injection Methods*

The first HI method was developed two and half decades ago for the synthesis of cadmium chalcogenide NCs. This approach is based on the rapid injection of a precursor into a boiling solvent [153]. Immediately after the injection, a rapid nucleation burst occurs with a simultaneous formation of small nuclei. A rapid depletion of monomers terminates the nucleation stage, after which the nuclei continue growing. Over time, this leads to the evolution of the population of the NCs, which is characterized by a narrow size distribution. This happens if the reaction is stopped when it is still in the size-focusing regime [154]. The key parameters to control the size, size-distribution, and shape of the colloidal NCs synthesized by the HI method are (i) the ratio of the surfactants to the precursors, (ii) the injection temperature of the cation or anion precursor, (iii) the reaction time, and (iv) the concentration of the precursors. In 2015, Protesescu *et al.* extended the HI approach to colloidal synthesis of cesium LHP NCs ($CsPbX_3$, X = Cl, Br, and I) [113]. $CsPbX_3$ NCs were obtained by injecting Cs-oleate into a hot solution (140–200°C) of PbX_2 (X = Cl, Br, I) salts, which served both as the Pb^{2+} and X^- source, dissolved in octadecene (ODE), carboxylic acids, and primary amines (Figure 2.19). Equal ratios of amines and acids resulted in the formation of monodisperse NCs whose size could be adjusted by varying the reaction temperature. Mixed-halide perovskite NCs could also be conveniently synthesized by simply adjusting the ratios of lead halide salts ($PbCl_2/PbBr_2$ or $PbBr_2/PbI_2$). The PL emission of the resulting NCs could be finely modulated across the entire visible spectrum (410–700 nm) by varying the halide composition or by tuning the size of NCs. Subsequently, the HI method was extended to $MAPbX_3$ (X = Br, I) NC systems by replacing Cs-oleate with a methylamine solution [145]. $MAPbBr_3$ and $MAPbI_3$ NCs were successfully obtained by varying the relative amount of OLA and OA capping ligands.

The HI protocol was further developed for the synthesis of perovskite-related lead halide-based materials. For instance, Cs_4PbX_6 (X = Cl, Br, I) NCs were successfully prepared by working under Cs^+ and OLA-rich conditions compared to those used in the conventional $CsPbBr_3$ NCs synthesis [155]. The resulting NCs were nearly monodisperse, and their size could be further tuned from 9 to 37 nm. Various other groups have reported the synthesis of $Cs_2Pb_2Br_5$ NCs using OAm and OA as surfactants in an excess of the $PbBr_2$ precursor [156].

In order to gain insight into the growth kinetics of $CsPbX_3$ NCs that are produced via the HI approach, Lignos *et al.* employed a droplet-based

(a)

(b)

CsPbBr₃- 8nm

100 nm

(c)

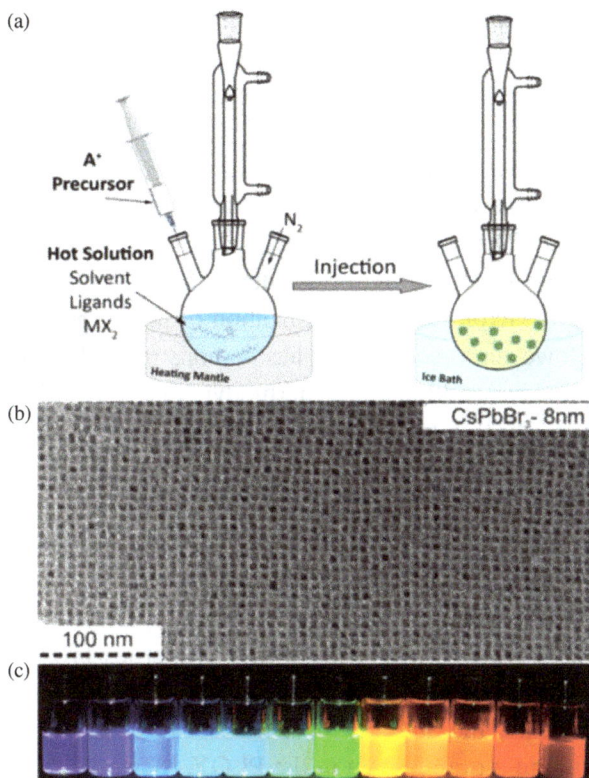

Figure 2.19. (a) Sketch of the HI method used for the synthesis of colloidal MHP NCs. (b) A typical TEM image of CsPbBr$_3$ NCs obtained using the hot injection (HI) strategy. (c) Colloidal perovskite CsPbX$_3$ (X = Cl, Br, I) NC dispersions (in each vial, the NCs have a different halide composition) in toluene under an ultraviolet (UV) lamp (λ = 365 nm). Reproduced from Ref. [113].

microfluidic platform [157,158]. By monitoring *in situ* the absorption and PL of CsPbX$_3$ NCs, the entire nucleation and growth processes are found to take place in the first 1–5 s of the reaction, suggesting extremely fast reaction kinetics. Slightly different results were reported by Koolyk *et al.*, who investigated the growth kinetics of CsPbBr$_3$ and CsPbI$_3$ NCs by taking aliquots at different stages of the reaction and analyzing them by transmission electron microscopy (TEM) [159]. The size focusing regime in CsPbI$_3$ NCs lasted for 20 s, followed by a subsequent size defocusing regime. However, the growth of $CsPbBr_3$ NCs was not characterized by any size focusing regime but by a broadening of the size distribution from the very beginning, which persisted during the whole reaction time of 40 s.

The fact that the nucleation and growth steps of MHP NCs are fast and hardly separable in time may be due to the highly ionic nature of the crystals formed. The challenge of controlling the nucleation and the growth with high structural uniformity has been one of the major obstacles impeding the exploration and utilization of the properties of $CsPbX_3$ NCs [160]. However, a recent study has demonstrated that this challenge can be overcome by working under thermodynamic equilibrium instead of kinetic control [161]. A major limitation of colloidal MHP NCs is their high solubility in polar solvents. This results in poor stability under ambient atmospheric conditions (with variable humidity, heat, and/or light). For example, perovskite NCs are particularly sensitive to the anti-solvents used in their cleaning [162]. For $CsPbBr_{1-x}I_x$ NCs, different anti-solvents, like isopropanol, *n*-butanol, or acetone, can change the halide composition of the NCs and, thus, their overall optical properties [162]. In order to counterbalance the stripping of halide ions, which accompanies the use of polar anti-solvents, the stability of $CsPbX_3$ NCs can be considerably enhanced, without a drastic drop in the PLQY, by working under halide-rich conditions [163]. In this case, $ZnBr_2$ was used as an extra source of Br ions in addition to the $PbBr_2$ precursor. The resulting NCs had more halide-rich compositions than those synthesized in the absence of metal bromide. This approach was considered as the first successful attempt to stabilize $CsPbX_3$ NCs (where X is Br or I) *in situ* by inorganic passivation. The term "inorganic passivation" is, however, not entirely accurate, as the NCs were still passivated with organic molecules. One major drawback of the HI methods is that they rely on the use of metal halide salts as both a cation and an anion precursor, and this limits the possibility of working with the desired ion stoichiometry [113,164].

In order to overcome such restrictions, Liu *et al.* developed the so-called "three-precursor" HI approach for the synthesis of $CsPbX_3$ (X = Cl, Br, or I) NCs [165]. The novelty consisted of using NH_4X (X = Cl, Br, or I) and PbO as sources of halide and lead ions separately, instead of conventional PbX_2 (X = Cl, Br, or I) salts. As had already been stated by

Woo *et al.* [163], the CsPbBr$_3$ NCs that had been synthesized under Br-rich conditions (by employing an excess of NH$_4$Br) had better optical properties and a remarkably better stability than the standard NCs, as they endured the purification step [165]. Yassitepe *et al.* further developed the three precursor HI approach in order to synthesize OA-capped CsPbX$_3$ NCs in the absence of alkylamines in the synthesis [166]. Cs-acetate and Pb-acetate were reacted with quaternary alkylammonium halides, such as tetraoctylammonium halides (TOA-X), which cannot form protonated ammonium species even when protons are present. The absence of OLA considerably speeds up the growth kinetics, enabling the synthesis of CsPbX$_3$ NCs at a lower temperature (i.e., 75°C). The CsPbBr$_3$ NCs that were obtained by this approach exhibited PLQY is up to 70%, as well as an enhanced colloidal stability. This method, however, failed to produce CsPbI$_3$ NCs of a similar quality and stability.

The three-precursors HI approach was later adopted and modified by Protesescu *et al.* for the colloidal synthesis of FAPbX$_3$ NCs [167,168]. FAPbBr$_3$ NCs were prepared by reacting FA and Pb acetates with OA in ODE and, subsequently, injecting oleylammonium bromide. The final product contained also a 5–10% of NH$_4$Pb$_2$Br$_5$ byproduct that might have formed upon the thermal decomposition of FA$^+$ to NH^{4+} during the synthesis [167]. Phase pure FAPbI$_3$ NCs with a better optical quality were successfully synthesized by the two-precursor HI method by reacting FA-oleate with a PbI$_2$ complex in the presence of OA and OLA in ODE with an excess of FA (FA: Pb = 2.7) at 80°C [169]. Although the three-precursors HI approach allows one to work with the desired stoichiometry of ions, its potential versatility is limited by a series of disadvantages. First, the fact that a synthesis of CsPbI$_3$, CsPbCl$_3$, and MAPbX$_3$ (X = Br, I) NCs has not yet been reported using this strategy suggests that the halide precursors react poorly under the reaction conditions of this approach [167,169,170]. Moreover, a considerable amount of undesired secondary phases, which are ascribed to the decomposition of the alkylammonium halide precursor, were found in the synthesis of FAPbX$_3$ NCs [167]. In order to compensate for the limitations of the three-precursor HI approach, new methods have recently been reported by Imran *et al.* [171] and by Creutz *et al.* [172]. Both these strategies are based on the use of highly reactive halides as anion precursors, which can be conveniently injected into a solution of metal carboxylates. Upon injection, the NCs immediately begin to nucleate and grow. Imran *et al.* demonstrated how benzoyl halides, which are employed as a halide source, can be used to prepare an entire family of all-inorganic and hybrid lead halide perovskite NCs (CsPbX$_3$, MAPbX$_3$, FAPbX$_3$; X = Cl⁻, Br⁻, I⁻) with good control over the size distribution and phase purity [171]. Of particular relevance is that these strategies enable one to work with desired cation/anion ratios, and, more specifically, in a halide-rich environment. In some systems, this was observed to strongly increase the PLQY of the resulting NCs: for instance,CsPbCl$_3$ NCs synthesized with either the LARP or the two-precursor HI approach were characterized by significant non-radiative carrier recombination [173,174], while the use of an excess of benzoyl chloride (halide-rich conditions) boosted their PLQY up to a record value of 65% [171].

2.3.4.1 *Size and shape control of nanocrystals by hot injection approach*

Size and shape control is made possible in the HI strategy mainly by varying the ligand combinations and ratios as well as the reaction temperature. As a general trend, the use of OLA and OA at low reaction temperatures (in the range of 90–130°C) tends to encourage the NCs to grow anisotropically, producing quasi-2D geometries, which are usually referred to as "nanoplatelets" (NPLs) [175]. On the other hand, high reaction temperatures (i.e., 170–200°C) and long reaction times lead to nanowires (NWs) [176]. In order to investigate the effect of ligands on the morphology of CsPbBr$_3$ NCs, Pan *et al.* carried out a comprehensive study by systematically varying the chain length of the alkyl amines and carboxylic acids used in the reaction [177]. In one series of experiments, while keeping the amount of OLA fixed, they added different carboxylic acids (at 170°C). An increase in the average edge length of CsPbBr$_3$ nanocubes, from 9.5 to 13 nm, occurred when the chain length of the carboxylic acids was shortened. Furthermore, working with OA and lowering the reaction temperature to 140°C, NPLs with a thickness of 2.5 nm and a width of 20 nm were formed (Figure 2.20). In a second series of experiments, the amount of OA was fixed, and different alkylamines were conducted at 170°C. In all experiments, NPLs were formed, except when OLA was

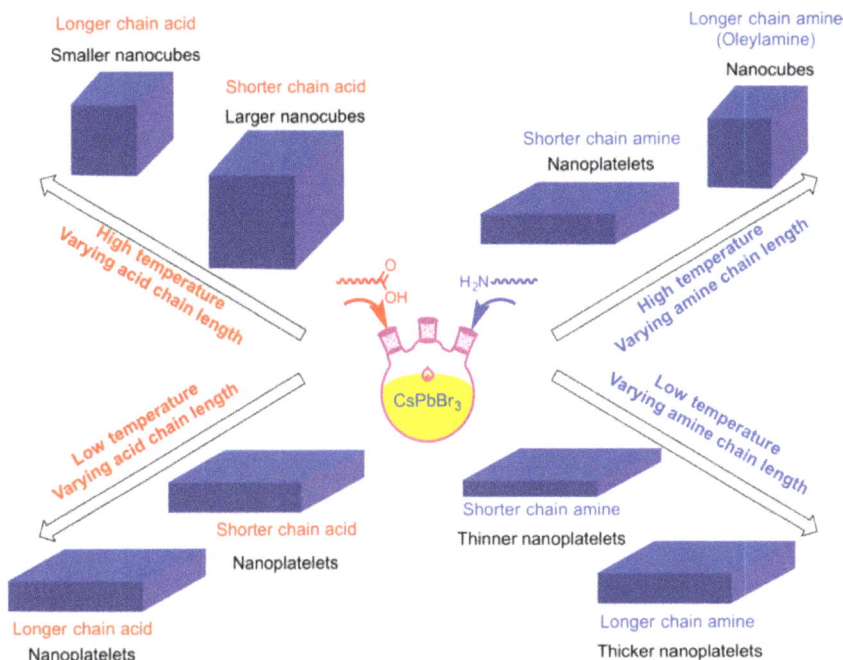

Figure 2.20. Schematic illustration of shape and size control of $CsPbBr_3$ NCs in the HI approach. Reproduced from Ref. [177].

used, which only leads to the formation of NPLs at lower reaction temperatures (140°C). While many attempts have been made to control the lateral size of lead halide-based nanocubes, some efforts have also been dedicated to synthesizing anisotropic nanostructures, such as NPLs and NWs, with controlled dimensions. For example, Song *et al.* reported a HI synthesis of atomically thin $CsPbBr_3$ nanosheets (NSs) with a thickness of 3.3 nm and edge length of about 1 μm [178]. Imran *et al.* demonstrated the synthesis of $CsPbBr_3$ NWs with a tunable width down to a few unit cells via the HI approach [179]. Green-emitting $CsPbBr_3$ NWs with a width of 10–20 nm were prepared by employing OAm and OLA only in the absence of carboxylic acid [180]. The diameter of the NWs could be decreased from 10 to 3.4 nm by introducing a short chain carboxylic acid (octanoic or hexanoic acid).

To better understand how alkylamines and carboxylic acids interact with each other before and during the HI synthesis, Almeida *et al.*

conducted an in-depth investigation into the interplay between OLA and OA, and studied the effects of the relative concentration on the size, size distribution, and shape of $CsPbBr_3$ NCs [161]. By increasing the concentration of ligands, the precipitation temperature of $PbBr_2$ could be significantly enhanced from 195 to 290°C, allowing syntheses of $CsPbBr_3$ NCs to be conducted at higher temperatures. Nuclear Magnetic Resonance (NMR) analysis indicates that the concentration of oleylammonium species could be increased not only by increasing the concentration of OA, but also by lowering the reaction temperature. The concentration of oleylammonium species was found to determine the shape of the final NCs. A high concentration of oleylammonium species, which is achievable in a highly acidic environment, causes the NCs to grow anisotropically, whereas a lower concentration leads to the formation of nanocubes. This demonstrates that oleylammonium, or primary alkyl ammonium in general, can compete with Cs^+ ions for the surface passivation of the NCs, generating platelet-shaped particles or even layered structures. It was also found that the Ostwald ripening could be suppressed by reducing the concentration of ligands. As a result, $CsPbBr_3$ nanocubes from 4.0 to 16.4 nm with narrow size distribution (8–15%) could be prepared. In addition to varying the length of the ligands and the ratio of amine to acid, control over the size of $CsPbBr_3$ NCs can also be achieved by employing extra halide sources such as alkylammonium bromide or $ZnBr_2$ salts [161,181]. The size of the $CsPbBr_3$ nanocubes could be finely tuned from 17.5 to 3.8 nm by varying the amount of OLA-HBr but keeping the reaction temperature and the ligand concentration fixed [181]. Dong *et al.* demonstrated excellent control over the size and size distribution of $CsPbX_3$ NCs by adjusting the reaction temperature and the ZnX_2/PbX_2 ratio in the reaction mixture [161]. It is particularly interesting that, in this latter case, an excess of ZnX_2 was found to strongly influence the surface passivation of the resulting NCs, eventually leading to a high PLQY.

A disadvantage of the HI method is that Cs/FA-oleate precursors, which have to be prepared in a separate flask before the actual reaction, are solid at room temperature and require a preheating step (up to 100°C) before their injection. It is not easy to up-scale this strategy since the injection of large quantities of a precursor at a high temperature results in a remarkable drop in the temperature as well as an inhomogeneous nucleation. Consequently, this method cannot be used for large scale production [113]. To overcome these limitations, alternative routes have been developed that rely on mixing all the chemicals together in one pot and letting them react either by using a heating mantle (heat-up or solvothermal approaches) or by ultrasonication or microwave irradiation [157,174,182–184]. For instance, Chen *et al.* reported a solvothermal synthesis of both $CsPbX_3$ nanocubes and NWs [185]. $CsPbX_3$ nanocubes were synthesized by directly mixing precursors, such as cesium carbonate and lead halide salts, together with ligands, and the resulting mixture was heated up in an autoclave at the desired temperature for a certain amount of time. Ultrathin $CsPbBr_3$ NWs were obtained when predissolved precursors, such as Cs-oleate and lead halide dissolved in ODE by using OA and OLA, were used. In 2016, Tong *et al.* reported a single step ultrasonication-assisted synthesis to produce $CsPbX_3$ NCs with a tunable halide composition, thickness, and morphology [174]. The procedure was extended later to longer reaction time in order to produce $CsPbX_3$ NWs [185]. Similarly, $CsPb(Br/I)_3$ nanorods (NRs) were also prepared by adjusting the ratios of ligands (OLA/OA) and the reaction temperature [186].

All approaches discussed so far mainly rely on binary ligands composed of carboxylic acids (mainly OA) and alkyl amines. The surface of LHPs NCs is dynamically stabilized with either oleylammonium halide or Cs oleate [187,188]. However, the final ligand composition strongly depends on the processing conditions. For instance, upon the use of polar solvents during washing cycles, the ammonium ligands are more prone to be detached from the surface than the carboxylate groups, and this eventually modifies the final PLQY [177]. To address this issue, Krieg *et al.* developed a new capping strategy based on long chain zwitter ionic molecules (i.e., 3- (*N,N*-dimethyloctadecylammonio) propanesulfonate) (Figure 2.21) [189]. These molecules bind strongly to the NC surface, which improves the chemical durability of the material. In particular, this class of ligands allows for the isolation of clean NCs with high PLQYs (above 90%) after four rounds of precipitation/redispersion, along with much higher overall reaction yields of uniform and colloidal dispersible NCs.

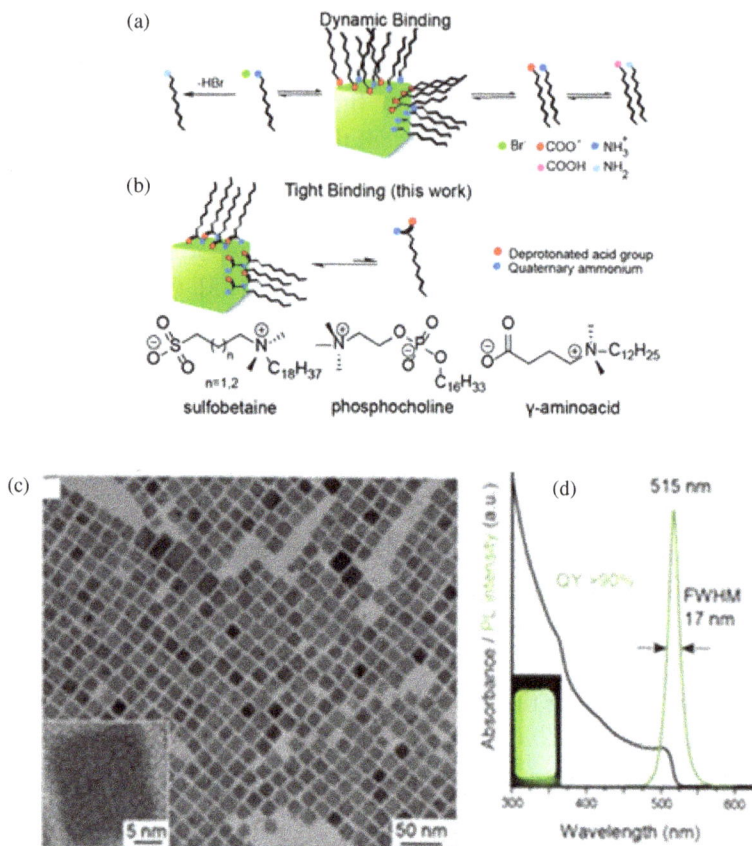

Figure 2.21.   (a) Depiction of conventional ligand capping of perovskite NCs using long-chain molecules with single head groups. (b) A novel strategy wherein cationic and anionic groups are combined in a single zwitterionic molecule. (c) Typical TEM images of zwitterionic-capped $CsPbBr_3$ NCs. (d) Absorbance and emission spectra of $CsPbBr_3$ NCs. Reproduced from Ref. [189].

2.3.4.2 *Mixed A/B Cations engineering of ABX_3 nanocrystals by hot injection*

Inspired by the tunable properties of LHP NCs with a mixed halide composition, efforts have been made to explore the possibility of preparing ABX_3 NCs using mixed A and B cations. For instance, in 2017, Amgar *et al.* reported the synthesis of $Cs_xRb_{1-x}BX_3$ by adjusting the Cs^+ and Rb^+

precursor ratios in the HI synthesis [190,191]. Interestingly, NC samples with a higher fraction of small Rb^+ ions had a larger bandgap. Similarly, the partial substitution of Pb^{2+} with Sn^{4+} ions enhanced the stability and optical properties of the NCs [192]. Protesescu *et al.* and Wang *et al.* reported an HI-based synthesis of mixed organic–inorganic $FA_xCs_{1-x}PbBr_{3-x}I_x$ and $FA_xCs_{x-1}PbI_3$ ($0 \leq x \leq 3$) perovskite NCs, NSs, and NWs [169,193]. It was concluded that the introduction of FA cations along with Cs ions in the A sites considerably enhances the stability of the NCs. Vashishtha *et al.* demonstrated that conventional monovalent A cations, such as Cs, Rb, MA, and FA, could be replaced by Tl^{3+} ions [194]. Tl_3PbX_5 NCs (X = Cl, Br, I) and $TlPbI_3$ NCs were prepared using the standard HI approach by replacing Cs-oleate with Tl-oleate [194]. The HI of Tl-oleate into the PbX_2 solution (130–175°C) resulted in faceted spheroidal Tl_3PbX_5 (X = Br, I) NCs with an orthorhombic crystal structure, whereas Tl_3PbCl_5 NCs crystallize in the tetragonal phase.

By making simple modifications to the standard HI approach, lead halide-based NC systems were successfully doped with either Mn^{2+}, Bi^{3+} or rare earth (RE) ions in order to alter their optical properties [195–200]. Liu *et al.* and Parobek *et al.* almost simultaneously reported an HI synthesis of Mn^{2+}-doped $CsPbCl_3$ NCs with fine control over the doping content [201,202]. In both cases, the incorporation of Mn^{2+} ions were achieved by simply employing $MnCl_2$, in addition to $PbCl_2$, in the conventional HI method that had previously been reported for the synthesis of $CsPbX_3$ NCs (Figures 2.22(a) and 2.22(b)). Later, Adhikari *et al.* further optimized the inclusion of Mn^{2+} ions in $CsPbCl_3$ NCs by using RNH_3Cl in addition to the $MnCl_2$ salt [203,204]. The authors claimed that the alkyl ammonium chloride precursor could allow the morphology to be precisely controlled and could enable Mn to be incorporated into $CsPbCl_3$ NCs. The following year, a similar strategy was adopted and slightly modified by Parobek *et al.* for the synthesis of Mn-doped $CsPbBr_3$ NCs [205]. Instead of using an alkyl ammonium halide precursor, the authors employed HBr, which was initially mixed with $PbBr_2$, $MnBr_2$, OA, and OLA to form $L_2[Pb_{1-x}Mnx]Br_4$ (L = ligand) organometallic complexes. Such compounds, which exhibit a strong Mn fluorescence, were subsequently transformed into Mn-doped $CsPbBr_3$ NCs by injecting the Cs precursor at a high temperature (Figure 2.22(c)).

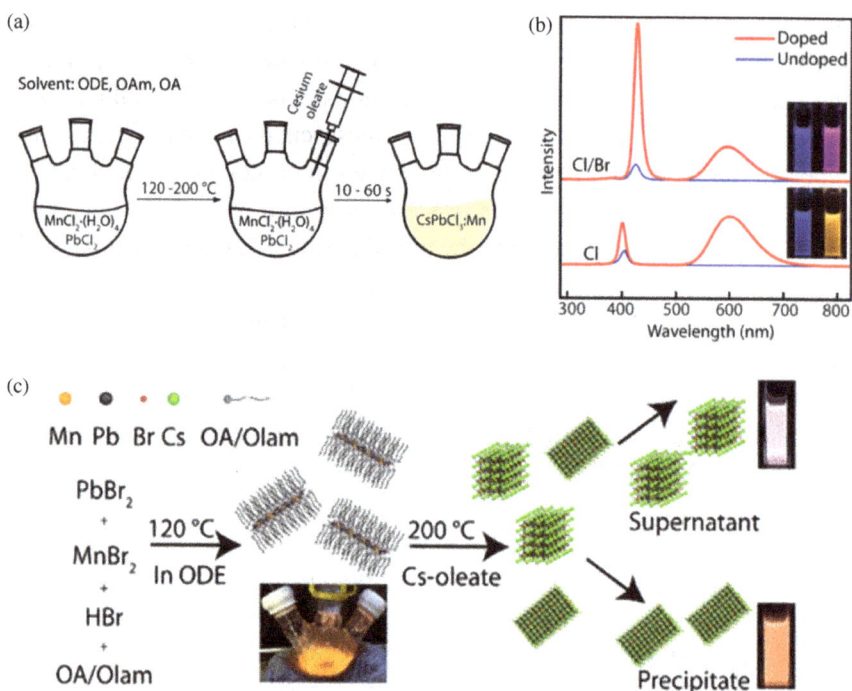

Figure 2.22.   Illustrations of the HI approaches used for the preparation of (a) Mn-doped $CsPbCl_3$, (b) Reproduced with permission from Ref. [202]. (c) Mn-doped $CsPbBr_3$ NCs. Reproduced from Ref. [205].

Interesting that doping $CsPbX_3$ NCs with Mn^{2+} ions increased their stability under ambient conditions [206,207]. Similarly, heterovalent dopants, such as Ce^{3+} and Bi^{3+} ions, were introduced into $CsPbBr_3$ NCs [208,209]. Milstein *et al.* recently reported ytterbium-doped $CsPbCl_3$ NCs by means of an HI synthesis following their previous three-precursors approach for double perovskites [210].

2.3.4.3 *Pb-Free MHP nanocrystals by hot injection*

The first colloidal synthesis of lead-free perovskite NCs via the HI approach was reported in 2016 by Jellicoe *et al.* [211], and Wang *et al.* [212] prepared $CsSnX_3$ NCs by reacting SnX_2 salts, which were dissolved in tri-n-octylphosphine (TOP), and injected the resulting solution into a mixture of $CsCO_3$, OA, and OLA at 170°C. Wang *et al.* fabricated Sn^{4+}-based perovskite NCs, with a Cs_2SnI_6 composition, which exhibited a PL peak around 620 nm (2.0 eV) and a full width at half-maximum of 49 nm (0.16 eV). The variation in the reaction time enabled the size and shape of Cs_2SnI_6 NCs to be tuned so that spherical QDs, NRs, NWs, nanobelts, and NPLs could be selectively prepared (Figure 2.23). The HI synthesis of Sn-based compounds was later modified slightly to access different shapes [212–214]. For example, Wang *et al.* synthesized 2D $CsSnI_3$ NPLs, with a thickness of less than 4 nm [213], by using a combination of long and short chain amines (OLA and OAm) and a short chain carboxylic acid (octanoic acid). In 2018, Wu *et al.* reported a synthesis of ternary $CsGeI_3$ NCs by means of an HI approach [215]. The NCs were synthesized by simply injecting Cs-oleate into a solution of GeI_2 which had been dissolved in ODE, OA, and OLA. Many attempts have been made to prepare Pb-free perovskite NCs by simply substituting Pb^{2+} ions with ternary cations, such as Bi^{3+} or Sb^{3+}, forming $Cs_3M_2X_9$ compounds or by the so-called

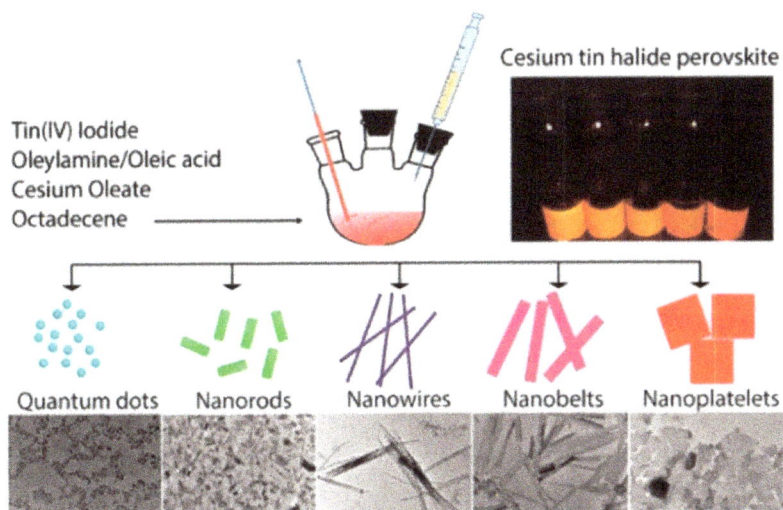

Figure 2.23. Scheme showing the synthesis of perovskite Cs_2SnI_6 NCs, with corresponding photographs of the as-prepared Cs_2SnI_6 samples under UV light and TEM images of Cs_2SnI_6 NCs with different shapes. Reproduced from Ref. [212].

"cation transmutation" strategy, namely, the replacement of two Pb^{2+} ions with one monovalent M^+ ion and one trivalent M^{3+} ion, forming a $A_2M^+M^{3+}X_6$ double-perovskite structure. Examples are $Cs_3Sb_2I_9$ and $Rb_3Sb_2I_9$ NCs that have been prepared with different morphologies using the HI synthesis route [216–218]. Also, in 2018, "double-perovskite" Cs_2AgBiX_6 (X = Cl or Br) NCs were synthesized by means of HI approaches [172,215,219,220]. Double perovskite NCs were reported by two groups using two different HI synthetic routes. Creutz *et al.* employed, for the first time, trimethylsilyl halides injected at 140°C into a solution of metal acetate precursors (i.e., silver acetate, cesium acetate, and bismuth acetate) [172]. The solution was then injected into ODE, OA, and OLA mixture solution, which immediately triggered the nucleation and growth of the NCs. Around the same time, Zhou *et al.* were the first to synthesize lead-free $Cs_2AgBiBr_6$ NCs by preparing a solution of $AgNO_3$, $BiBr_3$, ODE, OA, OLA, and HBr followed by injection of Cs-oleate at 200°C [219].

2.4 Alternative (Indirect) Synthesis Approaches

In addition to the synthesis approaches discussed above, there have been a few two-step synthesis approaches for generating LHP NCs in which the nucleation of LHPs takes place within starting colloidal seeds [155,221,222]. For example, presynthesized PbI_2 NCs were used as a colloidal template to form 2D and 3D LHP NCs when they were reacted with alkyl ammonium iodide or MAI [221]. Also, CsX NCs, which are easy to synthesize with tunable size, were employed as monodisperse colloidal precursors to make LHP NCs [222]. Interestingly, when CsBr NCs react with Pb-oleate, the CsBr to $CsPbBr_3$ transformation goes through $CsBr/CsPbBr_3$ core/shell NCs as intermediate structures. The ternary 0D compound (e.g., Cs_4PbBr_6 NCs) is another colloidal candidate that transforms into the LHP NCs (e.g., $CsPbBr_3$ NCs) when it is reacted with $PbBr_2$ (dissolved in toluene/OLA/OA) [155]. A potential advantage of such indirect synthesis is that control over the size of LHP NCs is possible by tuning the size of the parent NCs. Also, monodisperse colloidal NCs can, in principle, serve as precursors for LHP NCs with more complex morphologies.

2.5 Summary

The quality of the MHPs is the key to their applications, and the synthesis methodologies have a significant impact on the quality of the samples synthesized. This chapter provides an overview of the common synthesis techniques for the growth of bulk MHP SCs, MHP films, and NCs. The growth of bulk MHP SCs includes the STL, ITC, AVC, spin-coating, and CVD methods. The STL, ITC, and AVC methods are usually used to grow large-sized bulk MHP SCs. Although the STL method is more time-consuming, it allows the growth of high-quality large-scale bulk SCs. The ITC method not only depends on the temperature of the solution but also requires stronger solvents, such as DMF, DMSO, and GBA. The solvents are generally different depending on the specific halogen. Bulk MHP SCs can be grown rapidly by the AVC method, but proper anti-solvent is needed. Often two or three methods are used together for growing high-quality SCs. Growing high quality PSCs demands development of new techniques.

To prepare high-performance polycrystalline or microcrystalline MHP films, spin coating is commonly used. However, it has some limitations in the preparation of large-area films. The CVD method can be used to grow large-area MHP films by controlling the growth conditions. To obtain high-quality films, both spin-coating and CVD methods require independently controlling the nucleation and growth processes. Rapid nucleation of the perovskite (or intermediate phase) allows uniform coverage of the substrate and avoids the formation of detrimental microstructural defects such as voids and pinholes.

For the synthesis of NCs, LARP-related techniques are simple and convenient and allow for the synthesis of various $APbX_3$ NCs at room temperature with some control over the shape. However, the NCs made with such methods tend to have poor optical properties and stability due to the use of polar solvents in the synthesis process, which leads to degradation of the produced MHP NCs. The HI technique can be used to produce nearly monodisperse MHP NCs, with good control over the shape of the NCs. However, it needs to be performed in air-free conditions and is hard to scale up. Recently, many other strategies, such as the ultrasonic method, microwave-assisted method, solvothermal method. and ball milling, have also been developed to synthesize MHP NCs. By controlling the synthesis approaches, reaction temperature, capping ligands, and other reaction conditions, MHP NCs with different compositions (all inorganic or organic–inorganic hybrid), morphologies (0D, ID, 2D, 3D), and sizes/ shapes can be generated. Efforts have also been made to understand the related mechanisms of growth of the NCs.

References

[1] C. Zhang, X. Liu, J. Chen, and J. Lin, *Chinese J. Chem.* 39, 1353 (2021).
[2] F. Yao, J. Peng, R. Li, W. Li, P. Gui, B. Li, C. Liu, C. Tao, Q. Lin, and G. Fang, *Nat. Commun.* 11, 1194 (2020).
[3] G.M. Pound, M.T. Simnad, and L. Yang, *J. Chem. Phys.* 22, 1215 (1954).
[4] Y. Zhang, Y. Liu, Z. Xu, H. Ye, Q. Li, M. Hu, Z. Yang, and S. Liu, *J. Mater. Chem. C* 7, 1584 (2019).
[5] C.M. Raghavan, T.P. Chen, S.S. Li, W.L. Chen, C.Y. Lo, Y.M. Liao, G. Haider, C.C. Lin, C.C. Chen, R. Sankar, Y.M. Chan, F.C. Chou, and C.W. Chen, *Nano Lett.* 18, 3221 (2018).
[6] P. Nandi, C. Giri, D. Swain, U. Manju, and D. Topwal, *Crystengcomm* 21, 656 (2019).
[7] M. Bari, H. Wu, A.A. Bokov, R.F. Ali, H.N. Tailor, B.D. Gates, and Z.G. Ye, *Crystengcomm* 23, 3326 (2021).
[8] X. Liu, H. Zhang, B. Zhang, J. Dong, W. Jie, and Y. Xu, *J. Phys. Chem. C* 122, 14355 (2018).
[9] R. Babu, L. Giribabu, and S.P. Singh, *Cryst. Growth Des.* 18, 2645 (2018).
[10] D. Shi, V. Adinolfi, R. Comin, M. Yuan, E. Alarousu, A. Buin, Y. Chen, S. Hoogland, A. Rothenberger, K. Katsiev, Y. Losovyj, X. Zhang, P.A. Dowben, O.F. Mohammed, E.H. Sargent, and O.M. Bakr, *Science* 347, 519 (2015).
[11] C. Ge, M. Hu, P. Wu, Q. Tan, Z. Chen, Y. Wang, J. Shi, and J. Feng, *J. Phys. Chem. C* 122, 15973 (2018).
[12] J. Ding, S. Du, Z. Zuo, Y. Zhao, H. Cui, and X. Zhan, *J. Phys. Chem. C* 121, 4917 (2017).
[13] Y. Rakita, N. Kedem, S. Gupta, A. Sadhanala, V. Kalchenko, M.L. Böhm, M. Kulbak, R.H. Friend, D. Cahen, and G. Hodes, *Cryst. Growth Des.* 16, 5717 (2016).
[14] H. Zhang, X. Liu, J. Dong, H. Yu, C. Zhou, B. Zhang, Y. Xu, and W. Jie, *Cryst. Growth Des.* 17, 6426 (2017).

[15] M.I. Saidaminov, A.L. Abdelhady, B. Murali, E. Alarousu, V.M. Burlakov, W. Peng, I. Dursun, L. Wang, Y. He, G. Maculan, A. Goriely, T. Wu, O.F. Mohammed, and O.M. Bakr, *Nat. Commun.* 6, 7586 (2015).

[16] J.M. Kadro, K. Nonomura, D. Gachet, M. Gratzel, and A. Hagfeldt, *Sci. Rep.* 5, 11654 (2015).

[17] J. Ding, X. Cheng, L. Jing, T. Zhou, Y. Zhao, and S. Du, *ACS Appl. Mater. Interfaces* 10, 845 (2018).

[18] M.I. Saidaminov, A.L. Abdelhady, G. Maculan, and O.M. Bakr, *Chem. Commun. (Camb)* 51, 17658 (2015).

[19] M.I. Saidaminov, M.A. Haque, J. Almutlaq, S. Sarmah, X.-H. Miao, R. Begum, A.A. Zhumekenov, I. Dursun, N. Cho, B. Murali, O.F. Mohammed, T. Wu, and O.M. Bakr, *Adv. Opt. Mater.* 5, 1600704 (2017).

[20] Y. Liu, Y. Zhang, Z. Yang, J. Feng, Z. Xu, Q. Li, M. Hu, H. Ye, X. Zhang, M. Liu, K. Zhao, and S. Liu, *Mater. Today* 22, 67 (2019).

[21] Y. Liu, Y. Zhang, K. Zhao, Z. Yang, J. Feng, X. Zhang, K. Wang, L. Meng, H. Ye, M. Liu, and S.F. Liu, *Adv. Mater.* 30, 1707314 (2018).

[22] Z. Yuan, W. Huang, S. Ma, G. Ouyang, W. Hu, and W. Zhang, *J. Mater. Chem. C* 7, 5442 (2019).

[23] Y. Dang, Y. Liu, Y. Sun, D. Yuan, X. Liu, W. Lu, G. Liu, H. Xia, and X. Tao, *CrystEngComm.* 17, 665 (2015).

[24] Y.J. Fang, Q.F. Dong, Y.C. Shao, Y.B. Yuan, and J.S. Huang, *Nat. Photo.* 9, 679 (2015).

[25] E. Shi, Y. Gao, B.P. Finkenauer, Akriti, A.H. Coffey, and L. Dou, *Chem. Soc. Rev.* 47, 6046 (2018).

[26] J. Li, J. Wang, J. Ma, H. Shen, L. Li, X. Duan, and D. Li, *Nat. Commun.* 10, 806 (2019).

[27] H. Chen, J. Lin, J. Kang, Q. Kong, D. Lu, J. Kang, M. Lai, L.N. Quan, Z. Lin, J. Jin, L.-w. Wang, M.F. Toney, and P. Yang, *Sci. Adv.* 6, 4045 (2020).

[28] Y. Liu, H. Ye, Y. Zhang, K. Zhao, Z. Yang, Y. Yuan, H. Wu, G. Zhao, Z. Yang, J. Tang, Z. Xu, and S. Liu, *Matter* 1, 465 (2019).

[29] J. Shamsi, A.L. Abdelhady, S. Accornero, M. Arciniegas, L. Goldoni, A.R. Kandada, A. Petrozza, and L. Manna, *ACS Energy Lett.* 1, 1042 (2016).

[30] T. Yamada, T. Aharen, and Y. Kanemitsu, *Phys. Rev. Lett.* 120, 057404 (2018).

[31] K.H. Wang, L.C. Li, M. Shellaiah, and K. Wen Sun, *Sci. Rep.* 7, 13643 (2017).

[32] J. Ding, L. Jing, X. Cheng, Y. Zhao, S. Du, X. Zhan, and H. Cui, *J. Phys. Chem. Lett.* 9, 216 (2018).

[33] Y. Zhang, Y. Liu, Z. Yang, and S. Liu, *J. Energy Chem.* 27, 722 (2018).

[34] Z. Lian, Q. Yan, Q. Lv, Y. Wang, L. Liu, L. Zhang, S. Pan, Q. Li, L. Wang, and J.L. Sun, *Sci. Rep.* 5, 16563 (2015).

[35] Y. Dang, C. Zhong, G. Zhang, D. Ju, L. Wang, S. Xia, H. Xia, and X. Tao, *Chem. Mater.* 28, 6968 (2016).

[36] Y. Liu, Z. Yang, D. Cui, X. Ren, J. Sun, X. Liu, J. Zhang, Q. Wei, H. Fan, F. Yu, X. Zhang, C. Zhao, and S.F. Liu, *Adv. Mater.* 27, 5176 (2015).

[37] Y. Liu, J. Sun, Z. Yang, D. Yang, X. Ren, H. Xu, Z. Yang, and S.F. Liu, *Adv. Opt. Mater.* 4, 1829 (2016).

[38] Z. Lian, Q. Yan, T. Gao, J. Ding, Q. Lv, C. Ning, Q. Li, and J.L. Sun, *J. Am. Chem. Soc.* 138, 9409 (2016).

[39] C. Ge, W. Zhai, C. Tian, S. Zhao, T. Guo, S. Sun, W. Chen, and G. Ran, *RSC Adv.* 9, 16779 (2019).

[40] Y. Dang, Y. Zhou, X. Liu, D. Ju, S. Xia, H. Xia, and X. Tao, *Angewandte Chemie* 128, 3508 (2016).

[41] Y. Dang, G. Tong, W. Song, Z. Liu, L. Qiu, L. K. Ono, and Y. Qi, *J. Mater. Chem. C* 8, 276 (2020).

[42] Q. Dong, Y. Fang, Y. Shao, P. Mulligan, J. Qiu, L. Cao, and J. Huang, *Science* 347, 967 (2015).

[43] Y. Dong, Y. Zou, J. Song, X. Song, and H. Zeng, *J. Mater. Chem. C* 5, 11369 (2017).

[44] Y. Dang, D. Ju, L. Wang, and X. Tao, *CrystEngComm.* 18, 4476 (2016).

[45] I. Chung, B. Lee, J.Q. He, R.P.H. Chang, and M.G. Kanatzidis, *Nature* 485, 486 (2012).

[46] C.C. Stoumpos, C.D. Malliakas, J.A. Peters, Z. Liu, M. Sebastian, J. Im, T.C. Chasapis, A.C. Wibowo, D.Y. Chung, A.J. Freeman, B.W. Wessels, and M.G. Kanatzidis, *Cryst. Growth Des.* 13, 2722 (2013).

[47] J. Li, Q. Yu, Y. He, C.C. Stoumpos, G. Niu, G.G. Trimarchi, H. Guo, G. Dong, D. Wang, L. Wang, and M.G. Kanatzidis, *J. Am. Chem. Soc.* 140, 11085 (2018).

[48] M. Zhang, Z. Zheng, Q. Fu, Z. Chen, J. He, S. Zhang, C. Chen, and W. Luo, *J. Cryst. Growth* 484, 37 (2018).

[49] D. Zhou, L. Quan, X. Chen, S. Yu, Z. Zheng, and S. Gong, *J. Cryst. Growth* 311, 2524 (2009).

[50] M. Zhang, Z. Zheng, Q. Fu, Z. Chen, J. He, S. Zhang, L. Yan, Y. Hu, and W. Luo, *CrystEngComm.* 19, 6797 (2017).

[51] J. Song, Q. Cui, J. Li, J. Xu, Y. Wang, L. Xu, J. Xue, Y. Dong, T. Tian, H. Sun, and H. Zeng, *Adv. Optic. Mater.* 5, 1700157 (2017).

[52] Y. He, L. Matei, H.J. Jung, K.M. McCall, M. Chen, C.C. Stoumpos, Z. Liu, J.A. Peters, D.Y. Chung, B.W. Wessels, M.R. Wasielewski, V.P. Dravid, A. Burger, and M.G. Kanatzidis, *Nat. Commun.* 9, 1609 (2018).

[53] L. Braescu, *J. Colloid Interface Sci.* 319, 309 (2008).

[54] J. Lin, H. Chen, J. Kang, L.N. Quan, Z. Lin, Q. Kong, M. Lai, S. Yu, L. Wang, L.-w. Wang, M.F. Toney, and P. Yang, *Matter* 1, 180 (2019).

[55] G.C.B. Alexander, D.H. Fabini, R. Seshadri, and M.G. Kanatzidis, *Inorg. Chem.* 57, 804 (2018).

[56] M.A. Flatken, E. Radicchi, R. Wendt, A.G. Buzanich, E. Härk, J. Pascual, F. Mathies, O. Shargaieva, A. Prause, A. Dallmann, F. De Angelis, A. Hoell, and A. Abate, *Chem. Mater.* (2022).

[57] R. Ciesielski, F. Schäfer, N.F.Hartmann, N. Giesbrecht, T. Bein, P. Docampo, and A. Hartschuh, *ACS Appl. Mater. Interfaces* 10, 7974 (2018).

[58] Z. Xiao, R.A. Kerner, L. Zhao, N.L. Tran, K.M. Lee, T.-W. Koh, G.D. Scholes, and B.P. Rand, *Nat. Photo.* 11, 108 (2017).

[59] W.A. Dunlap-Shohl, Y. Zhou, N.P. Padture, and D.B. Mitzi, *Chem. Rev.* 119, 3193 (2019).

[60] N. Giesbrecht, J. Schlipf, L. Oesinghaus, A. Binek, T. Bein, P. Müller-Buschbaum, and P. Docampo, *ACS Energy Lett.* 1, 150 (2016).

[61] H. Duim, G.H. ten Brink, S. Adjokatse, R. de Kloe, B.J. Kooi, G. Portale, and M.A. Loi, *Small Structures* 1, 2000074 (2020).

[62] A.-F. Castro-Méndez, J. Hidalgo, and J.-P. Correa-Baena, *Adv. Energy Mater.* 9, 1901489 (2019).

[63] D. Kim, K. Higgins, and M. Ahmadi, *Matter* 4, 1442 (2021).

[64] J.-S. Park, J. Calbo, Y.-K. Jung, L.D. Whalley, and A. Walsh, *ACS Energy Lett.* 4, 1321 (2019).

[65] H. Mehdizadeh-Rad, F. Mehdizadeh-Rad, F. Zhu, and J. Singh, *Sol. Energy Mater. Sol. Cells* 220, 110837 (2021).

[66] L. Qiu, S. He, Y. Jiang, D.-Y. Son, L.K. Ono, Z. Liu, T. Kim, T. Bouloumis, S. Kazaoui, and Y. Qi, *J. Mater. Chem. A* 7, 6920 (2019).

[67] L. Qiu, S. He, Z. Liu, L.K. Ono, D.-Y. Son, Y. Liu, G. Tong, and Y. Qi, *J. Mater. Chem. A* 8, 23404 (2020).

[68] M. Liu, M.B. Johnston, and H.J. Snaith, *Nature* 501, 395 (2013).

[69] J.M. Ball, L. Buizza, H.C. Sansom, M.D. Farrar, M.T. Klug, J. Borchert, J. Patel, L.M. Herz, M.B. Johnston, and H.J. Snaith, *ACS Energy Lett.* 4, 2748 (2019).

[70] J.M. Ball, L. Buizza, H.C. Sansom, M. Farrar, and H.J. Snaith, *ACS Energy Lett.* 2019 (2019).

[71] C.-W. Chen, H.-W. Kang, S.-Y. Hsiao, P.-F. Yang, and K.-M. Chiang, *Adv. Mater.* 26, 6647 (2014).

[72] D. Zhou, T. Zhou, Y. Tian, X. Zhu, and Y. Tu, *J. Nanomater.* 2018, 8148072 (2018).

[73] C.-W. Chen, H.-W. Kang, S.-Y. Hsiao, P.-F. Yang, K.-M. Chiang, and H.-W. Lin, *Adv. Mater.* 26, 6647 (2014).

[74] R. Guo, D. Han, W. Chen, L. Dai, K. Ji, Q. Xiong, S. Li, L.K. Reb, M.A. Scheel, S. Pratap, N. Li, S. Yin, T. Xiao, S. Liang, A.L. Oechsle, C.L. Weindl, M. Schwartzkopf, H. Ebert, P. Gao, K. Wang, M. Yuan, N.C. Greenham, S.D. Stranks, S.V. Roth, R.H. Friend, and P. Müller-Buschbaum, *Nat. Energy* 6, 977 (2021).

[75] A. Kojima, K. Teshima, Y. Shirai, and T. Miyasaka, *J. Am. Chem. Soc.* 131, 6050 (2009).

[76] J. Pan, C. Mu, Q. Li, W. Li, D. Ma, and D. Xu, *Adv. Mater.* 28, 8309 (2016).

[77] G. Tang, P. You, Q. Tai, R. Wu, and F. Yan, *Solar RRL* 2, 1800066 (2018).

[78] G. Tang, P. You, Q. Tai, R. Wu, and F. Yan, *Solar RRL* 2 (2018).

[79] L. Zhao, D. Luo, J. Wu, Q. Hu, W. Zhang, K. Chen, T. Liu, Y. Liu, Y. Zhang, F. Liu, T.P. Russell, H.J. Snaith, R. Zhu, and Q. Gong, *Adv. Funct. Mater.* 26, 3508 (2016).

[80] Y. Wu, F. Xie, H. Chen, X. Yang, H. Su, M. Cai, Z. Zhou, T. Noda, and L. Han, *Adv. Mater.* 29, 1701073 (2017).

[81] Y. Miao, X. Liu, Y. Chen, T. Zhang, T. Wang, and Y. Zhao, *Adv. Mater.* 33, 2105699 (2021).

[82] B.-w. Park, H.W. Kwon, Y. Lee, D.Y. Lee, M.G. Kim, G. Kim, K.-j. Kim, Y.K. Kim, J. Im, T.J. Shin, and S.I. Seok, *Nat. Energy* 6, 419 (2021).

[83] M. Jung, S.-G. Ji, G. Kim, and S.I. Seok, *Chem. Soc Rev.* 48, 2011 (2019).

[84] D. Shen, X. Yu, X. Cai, M. Peng, Y. Ma, X. Su, L. Xiao, and D. Zou, *J. Mater. Chem. A* 2, 20454 (2014).

[85] J. Nam, J. Noh, J. Hong, K. Young, and C. Yang, *Nat. Materials* (2014).

[86] L. Chao, T. Niu, H. Gu, Y. Yang, Q. Wei, Y. Xia, W. Hui, S. Zuo, Z. Zhu, C. Pei, X. Li, J. Zhang, J. Fang, G. Xing, H. Li, X. Huang, X. Gao, C. Ran, L. Song, L. Fu, Y. Chen, and W. Huang, *Research* 2020, 2616345 (2020).

[87] H. Zhang, C. Tu, C. Xue, J. Wu, Y. Cao, W. Zou, W. Xu, K. Wen, J. Zhang, Y. Chen, J. Lai, L. Zhu, K. Pan, L. Xu, Y. Wei, H. Lin, N. Wang, W. Huang, and J. Wang, *Nano Lett.* 21, 3738 (2021).

[88] J. Burschka, N. Pellet, S.-J. Moon, R. Humphry-Baker, P. Gao, M.K. Nazeeruddin, and M. Grätzel, *Nature* 499, 316 (2013).

[89] L. Zheng, Y. Ma, S. Chu, S. Wang, B. Qu, L. Xiao, Z. Chen, Q. Gong, Z. Wu, and X. Hou, *Nanoscale* 6, 8171 (2014).

[90] Z. Xiao, C. Bi, Y. Shao, Q. Dong, Q. Wang, Y. Yuan, C. Wang, Y. Gao, and J. Huang, *Energy Environ. Sci.* 7, 2619 (2014).

[91] J.-H. Im, H.-S. Kim, and N.-G. Park, *APL Mater.* 2, 081510 (2014).

[92] Q. Chen, H. Zhou, Z. Hong, S. Luo, H.-S. Duan, H.-H. Wang, Y. Liu, G. Li, and Y. Yang, *J. Am. Chem. Soc.* 136, 622 (2014).

[93] Q. Chen, H. Zhou, Z. Hong, S. Luo, H.S. Duan, H.H. Wang, Y. Liu, G. Li, Y. Yang, *J Am Chem Soc* 136, 622 (2014).

[94] P. Luo, Z. Liu, W. Xia, C. Yuan, J. Cheng, and Y. Lu, *ACS Appl. Mater. Interfaces* 7, 2708 (2015).

[95] M.R. Leyden, M.V. Lee, S.R. Raga, and Y. Qi, *J. Mater. Chem. A* 3, 16097 (2015).

[96] P. Luo, Z. Liu, W. Xia, C. Yuan, J. Cheng, and Y. Lu, *J. Mater. Chem. A* 3, 12443 (2015).

[97] M.-H. Li, H.-H. Yeh, Y.-H. Chiang, U.-S. Jeng, C.-J. Su, H.-W. Shiu, Y.-J. Hsu, N. Kosugi, T. Ohigashi, Y.-A. Chen, P.-S. Shen, P. Chen, and T.-F. Guo, *Adv. Mater.* 30, 1801401 (2018).

[98] J. Duan, Y. Zhao, B. He, and Q. Tang, *Angewandte Chemie International Edition* 57, 3787 (2018).

[99] J. Duan, Y. Zhao, B. He, and Q. Tang, *Small* 14, 1704443 (2018).

[100] R. Liu, J. Zhang, H. Zhou, Z. Song, Z. Song, C.R. Grice, D. Wu, L. Shen, and H. Wang, *Adv. Opt. Mater.* 8, 1901735 (2020).

[101] M.F. Achoi, M.A.A. Noman, S. Kato, N. Kishi, and T. Soga, *J. Electron. Mater.* 51, 577 (2022).

[102] Y. Deng, E. Peng, Y. Shao, Z. Xiao, Q. Dong, and J. Huang, *Energy Environ. Sci.* 8, 1544 (2015).

[103] M. Du, X. Zhu, L. Wang, H. Wang, J. Feng, X. Jiang, Y. Cao, Y. Sun, L. Duan, Y. Jiao, K. Wang, X. Ren, Z. Yan, S. Pang, and S. Liu, *Adv. Mater.* 32, 2004979 (2020).

[104] T.R. Andersen, H.F. Dam, M. Hösel, M. Helgesen, J.E. Carlé, T.T. Larsen-Olsen, S.A. Gevorgyan, J.W. Andreasen, J. Adams, N. Li, F. Machui, G.D. Spyropoulos, T. Ameri, N. Lemaître, M. Legros, A. Scheel, D. Gaiser, K. Kreul, S. Berny, O.R. Lozman, S. Nordman, M. Välimäki, M. Vilkman, R.R. Søndergaard, M. Jørgensen, C.J. Brabec, and F.C. Krebs, *Energy Environ. Sci.* 7, 2925 (2014).

[105] Z. Wei, H. Chen, K. Yan, and S. Yang, *Angew. Chem. Int. Ed. Engl.* 53, 13239 (2014).

[106] L.C. Schmidt, A. Pertegas, S. Gonzalez-Carrero, O. Malinkiewicz, S. Agouram, G. Minguez Espallargas, H.J. Bolink, R.E. Galian, J. Perez-Prieto, *J. Am. Chem. Soc.* 136, 850 (2014).

[107] J. Shamsi, A.S. Urban, M. Imran, L. De Trizio, and L. Manna, *Chem. Rev.* 119, 3296 (2019).

[108] H. Huang, L. Polavarapu, J.A. Sichert, A.S. Susha, A.S. Urban, A.L. Rogach, *NPG Asia Mater.* 8, e328 (2016).

[109] J. Wang, J.L. Liu, Z.L. Du, and Z.Q. Li, *J. Energy Chem.* 54, 770 (2021).

[110] C.H. Lu, G.V. Biesold-McGee, Y. Liu, Z. Kang, and Z. Lin, *Chem. Soc. Rev.* 49, 4953 (2020).

[111] Y. Li, X. Zhang, H. Huang, S.V. Kershaw, and A.L. Rogach, *Mater. Today* 32, 204 (2020).

[112] F. Zhang, H. Zhong, C. Chen, X.G. Wu, X. Hu, H. Huang, J. Han, B. Zou, and Y. Dong, *ACS Nano* 9, 4533 (2015).

[113] L. Protesescu, S. Yakunin, M.I. Bodnarchuk, F. Krieg, R. Caputo, C.H. Hendon, R.X. Yang, A. Walsh, and M.V. Kovalenko, *Nano Lett.* 15, 3692 (2015).

[114] I. Levchuk, A. Osvet, X. Tang, M. Brandl, J.D. Perea, F. Hoegl, G.J. Matt, R. Hock, M. Batentschuk, and C.J. Brabec, *Nano Lett.* 17, 2765 (2017).

[115] G.C. Papavassiliou, G. Pagona, N. Karousis, G.A. Mousdis, I. Koutselas, and A. Vassilakopoulou, *J. Mater. Chem.* 22, 8271 (2012).

[116] H. Huang, A.S. Susha, S.V. Kershaw, T.F. Hung, and A.L. Rogach, *Adv. Sci. (Weinh)* 2, 1500194 (2015).

[117] P. Arunkumar, K.H. Gil, S. Won, S. Unithrattil, Y.H. Kim, H.J. Kim, and W.B. Im, *J. Phys. Chem. Lett.* 8, 4161 (2017).

[118] Y. Zhao, X. Xu, and X. You, *Sci. Rep.* 6, 35931 (2016).

[119] B. Luo, Y.C. Pu, S.A. Lindley, Y. Yi, and J.Z. Zhang, *Angew. Chemie. Int. Ed.* 128 (2016).

[120] S.A. Veldhuis, Y.K.E. Tay, A. Bruno, S.S.H. Dintakurti, S. Bhaumik, S.K. Muduli, M. Li, N. Mathews, T.C. Sum, and S.G. Mhaisalkar, *Nano Lett.* 17, 7424 (2017).

[121] G.H. Ahmed, J. Yin, R. Bose, L. Sinatra, E. Alarousu, E. Yengel, N.M. AlYami, M.I. Saidaminov, Y. Zhang, M.N. Hedhili, O.M. Bakr, J.-L. Brédas, and O.F. Mohammed, *Chem. Mater.* 29, 4393 (2017).

[122] B. Luo, S.B. Naghadeh, A.L. Allen, X. Li, and J.Z. Zhang, *Adv. Funct. Mater.* 27, 1604018 (2017).

[123] S. Wang, L. Zhou, F. Huang, Y. Xin, P. Jin, Q. Ma, Q. Pang, Y. Chen, and J.Z. Zhang, *J. Mater. Chem. C* 6, 10994 (2018).

[124] S.W. Dai, B.W. Hsu, C.Y. Chen, C.A. Lee, H. Y. Liu, H.F. Wang, Y.C. Huang, T.L. Wu, A. Manikandan, and R.M. Ho, *Adv. Mater.* 30, 1870048 (2018).

[125] G.C. Papavassiliou and I.B. Koutselas, *Synth. Met.* 71, 1713 (1995).

[126] S. Kumar, J. Jagielski, S. Yakunin, P. Rice, Y.C. Chiu, M.C. Wang, G. Nedelcu, Y. Kim, S.C. Lin, E.J.G. Santos, M.V. Kovalenko, and C.J. Shih, *ACS Nano* 10, 9720 (2016).

[127] J. Cho, Y.-H. Choi, T.E.O'Loughlin, L. De Jesus, and S. Banerjee, *Chem. Mater.* 28, 6909 (2016).

[128] I. Levchuk, P. Herre, M. Brandl, A. Osvet, R. Hock, W. Peukert, P. Schweizer, E. Spiecker, M. Batentschuk, and C.J. Brabec, *Chem. Commun. (Camb)* 53, 244 (2016).

[129] H. Huang, J. Raith, S.V. Kershaw, S. Kalytchuk, O. Tomanec, L. Jing, A.S. Susha, R. Zboril, and A.L. Rogach, *Nat. Commun.* 8, 996 (2017).

[130] M.C. Weidman, M. Seitz, S.D. Stranks, and W.A. Tisdale, *ACS Nano* 10, 7830 (2016).

[131] A. Perumal, S. Shendre, M. Li, Y.K. Tay, V.K. Sharma, S. Chen, Z. Wei, Q. Liu, Y. Gao, P.J. Buenconsejo, S.T. Tan, C.L. Gan, Q. Xiong, T.C. Sum, and H.V. Demir, *Sci. Rep.* 6, 36733 (2016).

[132] D.N. Minh, J. Kim, J. Hyon, J.H. Sim, H.H. Sowlih, C. Seo, J. Nam, S. Eom, S. Suk, S. Lee, E. Kim, and Y. Kang, *Chem. Mater.* 29, 5713 (2017).

[133] S. Kumar, J. Jagielski, N. Kallikounis, Y.H. Kim, C. Wolf, F. Jenny, T. Tian, C.J. Hofer, Y.C. Chiu, W.J. Stark, T.W. Lee, and C.J. Shih, *Nano Lett.* 17, 5277 (2017).

[134] X. Li, Y. Wu, S. Zhang, B. Cai, Y. Gu, J. Song, and H. Zeng, *Adv. Funct. Mater.* 26, 2435 (2016).

[135] S.T. Ha, R. Su, J. Xing, Q. Zhang, and Q. Xiong, *Chem. Sci.* 8, 2522 (2017).

[136] S. Seth and A. Samanta, *Sci. Rep.* 6, 37693 (2016).

[137] A. Kostopoulou, M. Sygletou, K. Brintakis, A. Lappas, and E. Stratakis, *Nanoscale* 10.1039.C7NR06404G (2017).

[138] X. Zhang, X. Bai, H. Wu, X. Zhang, C. Sun, Y. Zhang, W. Zhang, W. Zheng, W.W. Yu, and A.L. Rogach, *Angewandte Chemie* (2018).

[139] G.E. Eperon, C.E. Beck, and H.J. Snaith, *Mater. Horiz.* 3, 63 (2016).

[140] M. Mittal, A. Jana, S. Sarkar, P. Mahadevan, and S. Sapra, *J. Phys. Chem. Lett.* 7, 3270 (2016).

[141] D. Chen, X. Chen, Z. Wan, and G. Fang, *ACS Appl. Mater. Interfaces* 9, 20671 (2017).

[142]   X. Zhang, H. Liu, W. Wang, J. Zhang, B. Xu, K.L. Karen, Y. Zheng, S. Liu, S. Chen, K. Wang, and X.W. Sun, *Adv. Mater.* 29 (2017).

[143]   B. Xu, W. Wang, X. Zhang, W. Cao, D. Wu, S. Liu, H. Dai, S. Chen, K. Wang, and X. Sun, *J. Mater. Chem. C* 5, 6123 (2017).

[144]   Y.-W. Zhang, G. Wu, H. Dang, K. Ma, and S. Chen, *Indust. Eng. Chem. Res.* 56, 10053 (2017).

[145]   O. Vybornyi, S. Yakunin, and M.V. Kovalenko, *Nanoscale* 8, 6278 (2016).

[146]   L. Ruan, W. Shen, A. Wang, A. Xiang, and Z. Deng, *J. Phys. Chem. Lett.* 8, 3853 (2017).

[147]   L. Ruan, J. Lin, W. Shen, and Z. Deng, *Nanoscale* 10.1039.C8NR00883C (2018).

[148]   M. Leng, Z. Chen, Y. Yang, Z. Li, K. Zeng, K. Li, G. Niu, Y. He, Q. Zhou, and J. Tang, *Angew. Chem. Int. Ed. Engl.* 55, 15012 (2016).

[149]   M. Leng, Y. Yang, K. Zeng, Z. Chen, Z. Tan, S. Li, J. Li, B. Xu, D. Li, M. P. Hautzinger, Y. Fu, T. Zhai, L. Xu, G. Niu, S. Jin, and J. Tang, *Adv. Funct. Mater.* 28, 1704446 (2017).

[150]   J. Zhang, Y. Yang, H. Deng, U. Farooq, X. Yang, J. Khan, J. Tang, and H. Song, *ACS Nano* 11, 9294 (2017).

[151]   X. Li, D. Yu, F. Cao, Y. Gu, Y. Wei, Y. Wu, J. Song, and H. Zeng, *Adv. Funct. Mater.* 26, 5903 (2016).

[152]   F. Zhang, S. Huang, P. Wang, X. Chen, S. Zhao, Y. Dong, and H. Zhong, *Chem. Mater.* 29, 3793 (2017).

[153]   W. And and X. Peng, *Angew. Chemie.* (2002).

[154]   L. Manna, D.J. Milliron, A. Meisel, E.C. Scher, and A.P. Alivisatos, *Nat. Mater.* 2, 382 (2003).

[155]   Q.A. Akkerman, S. Park, E. Radicchi, F. Nunzi, E. Mosconi, F. De Angelis, R. Brescia, P. Rastogi, M. Prato, and L. Manna, *Nano Lett.* 17, 1924 (2017).

[156]   C. Han, C. Li, Z. Zang, M. Wang, K. Sun, X. Tang, and J. Du, *Photo. Res.* 5, 473 (2017).

[157]   I. Lignos, S. Stavrakis, G. Nedelcu, L. Protesescu, A.J. deMello, and M.V. Kovalenko, *Nano Lett.* 16, 1869 (2016).

[158]   R.M. Maceiczyk, K. Dümbgen, I. Lignos, L. Protesescu, M.V. Kovalenko, and A.J. deMello, *Chem. Mater.* 29, 8433 (2017).

[159]   M. Koolyk, D. Amgar, S. Aharon, and L. Etgar, *Nanoscale* 8, 6403 (2016).

[160]   G. Almeida, L. Goldoni, Q. Akkerman, Z. Dang, A.H. Khan, S. Marras, I. Moreels, and L. Manna, *ACS Nano* 12, 1704 (2018).

[161] Y. Dong, T. Qiao, D. Kim, D. Parobek, D. Rossi, and D.H. Son, *Nano Lett.* 18, 3716 (2018).

[162] L. Yuan, R. Patterson, X. Wen, Z. Zhang, G. Conibeer, and S. Huang, *J. Colloid Interface Sci* 504, 586 (2017).

[163] J.Y. Woo, Y. Kim, J. Bae, T.G. Kim, J.W. Kim, D.C. Lee, S. Jeong, *Chem. Mater.* 29, 7088 (2017).

[164] Y. Su, X. Chen, W. Ji, Q. Zeng, Z. Ren, Z. Su, and L. Liu, *ACS Appl. Mater. Interfaces* 9, 33020 (2017).

[165] P. Liu, W. Chen, W. Wang, B. Xu, D. Wu, J. Hao, W. Cao, F. Fang, Y. Li, Y. Zeng, R. Pan, S. Chen, W. Cao, X.W. Sun, and K. Wang, *Chem. Mater.* 29, 5168 (2017).

[166] E. Yassitepe, Z. Yang, O. Voznyy, Y. Kim, G. Walters, J.A. Castañeda, P. Kanjanaboos, M. Yuan, X. Gong, F. Fan, J. Pan, S. Hoogland, R. Comin, O. M. Bakr, L.A. Padilha, A.F. Nogueira, and E.H. Sargent, *Adv. Funct. Mater.* 26, 8757 (2016).

[167] L. Protesescu, S. Yakunin, M.I. Bodnarchuk, F. Bertolotti, N. Masciocchi, A. Guagliardi, and M.V. Kovalenko, *J. Am. Chem. Soc.* 138, 14202 (2016).

[168] I. Lignos, L. Protesescu, D.B. Emiroglu, R. Maceiczyk, S. Schneider, M.V. Kovalenko, and A.J. deMello, *Nano Lett.* 18, 1246 (2018).

[169] L. Protesescu, S. Yakunin, S. Kumar, J. Bar, F. Bertolotti, N. Masciocchi, A. Guagliardi, M. Grotevent, I. Shorubalko, M.I. Bodnarchuk, C.J. Shih, and M.V. Kovalenko, *ACS Nano* 11, 3119 (2017).

[170] Q. Li, H. Li, H. Shen, F. Wang, F. Zhao, F. Li, X. Zhang, D. Li, X. Jin, and W. Sun, *ACS Photo.* 4, 2504 (2017).

[171] M. Imran, V. Caligiuri, M.J. Wang, L. Goldoni, M. Prato, R. Krahne, L. De Trizio, and L. Manna, *J. Am. Chem. Soc.* 140, 2656 (2018).

[172] S.E. Creutz, E.N. Crites, M.C. De Siena, and D.R. Gamelin, *Nano Lett.* 18, 1118 (2018).

[173] Y. Kim, E. Yassitepe, O. Voznyy, R. Comin, G. Walters, X. Gong, P. Kanjanaboos, A.F. Nogueira, and E.H. Sargent, *ACS Appl. Mater. Interfaces* 7, 25007 (2015).

[174] Y. Tong, E. Bladt, M.F. Ayguler, A. Manzi, K.Z. Milowska, V.A. Hintermayr, P. Docampo, S. Bals, A.S. Urban, L. Polavarapu, and J. Feldmann, *Angew. Chem. Int. Ed. Engl.* 55, 13887 (2016).

[175] Y. Bekenstein, B.A. Koscher, S.W. Eaton, P.D. Yang, and A.P. Alivisatos, *J. Am. Chem. Soc.* 137, 16008 (2015).

[176] D. Zhang, S.W. Eaton, Y. Yu, L. Dou, and P. Yang, *J. Am. Chem. Soc.* 137, 9230 (2015).

[177] A. Pan, B. He, X. Fan, Z. Liu, J.J. Urban, A.P. Alivisatos, L. He, and Y. Liu, *ACS Nano* 10, 7943 (2016).

[178] J. Song, L. Xu, J. Li, J. Xue, Y. Dong, X. Li, and H. Zeng, *Adv. Mater.* 28, 4861 (2016).

[179] M. Imran, F. Di Stasio, Z. Dang, C. Canale, A.H. Khan, J. Shamsi, R. Brescia, M. Prato, and L. Manna, *Chem. Mater.* 28, 6450 (2016).

[180] D. Zhang, Y. Yang, Y. Bekenstein, Y. Yu, N.A. Gibson, A.B. Wong, S.W. Eaton, N. Kornienko, Q. Kong, M. Lai, A.P. Alivisatos, S.R. Leone, and P. Yang, *J. Am. Chem. Soc.* 138, 7236 (2016).

[181] A. Dutta, S.K. Dutta, S. Das Adhikari, and N. Pradhan, *ACS Energy Lett.* 3, 329 (2018).

[182] Q. Pan, H. Hu, Y. Zou, M. Chen, L. Wu, D. Yang, X. Yuan, J. Fan, B. Sun, and Q. Zhang, *J. Mater. Chem. C* 5, 10947 (2017).

[183] D.M. Jang, D.H. Kim, K. Park, J. Park, J.W. Lee, and J.K. Song, *J. Mater. Chem. C* 4, 10625 (2016).

[184] Z. Long, H. Ren, J.H. Sun, J. Ouyang, and N. Na, *Chem. Commun.* 53, 9914 (2017).

[185] M. Chen, Y. Zou, L. Wu, Q. Pan, D. Yang, H. Hu, Y. Tan, Q. Zhong, Y. Xu, H. Liu, B. Sun, and Q. Zhang, *Adv. Funct. Mater.* 27, 1701121 (2017).

[186] X. Tang, Z. Zu, H. Shao, W. Hu, M. Zhou, M. Deng, W. Chen, Z. Zang, T. Zhu, and J. Xue, *Nanoscale* 8, 15158 (2016).

[187] Z. Dang, J. Shamsi, F. Palazon, M. Imran, Q.A. Akkerman, S. Park, G. Bertoni, M. Prato, R. Brescia, and L. Manna, *ACS Nano* 11, 2124 (2017).

[188] G.G. Huang, C.L. Wang, S.H. Xu, S.F. Zong, J. Lu, Z.Y. Wang, C.G. Lu, and Y.P. Cui, *Adv. Mater.* 29 (2017).

[189] F. Krieg, S.T. Ochsenbein, S. Yakunin, S. Ten Brinck, P. Aellen, A. Suess, B. Clerc, D. Guggisberg, O. Nazarenko, Y. Shynkarenko, S. Kumar, C.J. Shih, I. Infante, and M.V. Kovalenko, *ACS Energy Lett.* 3, 641 (2018).

[190] D. Amgar, T. Binyamin, V. Uvarov, and L. Etgar, *Nanoscale* 10, 6060 (2018).

[191] H. Wu, Y. Yang, D. Zhou, K. Li, J. Yu, J. Han, Z. Li, Z. Long, J. Ma, and J. Qiu, *Nanoscale* 10, 3429 (2018).

[192] H.C. Wang, W. Wang, A.C. Tang, H.Y. Tsai, Z. Bao, T. Ihara, N. Yarita, H. Tahara, Y. Kanemitsu, and S. Chen, *Angew. Chemie Int. Ed.* (2017).

[193] C. Wang, Y. Zhang, A. Wang, Q. Wang, H. Tang, W. Shen, Z. Li, and Z. Deng, *Chem. Mater.* 29, 2157 (2017).

[194] P. Vashishtha, D.Z. Metin, M.E. Cryer, K. Chen, J.M. Hodgkiss, N. Gaston, and J.E. Halpert, *Chem. Mater.* 30, 2973 (2018).

[195] G. Pan, X. Bai, D. Yang, X. Chen, P. Jing, S. Qu, L. Zhang, D. Zhou, J. Zhu, W. Xu, B. Dong, and H. Song, *Nano Lett.* 17, 8005 (2017).

[196] Q.S. Hu, Z. Li, Z. F. Tan, H.B. Song, C. Ge, G.D. Niu, J.T. Han, and J. Tang, *Adv. Opt. Mater.* 6 (2018).

[197] A. Biswas, R. Bakthavatsalam, and J. Kundu, *Chem. Mater.* 29, 7816 (2017).

[198] P. Wang, B. Dong, Z. Cui, R. Gao, G. Su, W. Wang, and L. Cao, *RSC Adv.* 8, 1940 (2018).

[199] H. Shao, X. Bai, H. Cui, G. Pan, P. Jing, S. Qu, J. Zhu, Y. Zhai, B. Dong, and H. Song, *Nanoscale* 10, 1023 (2018).

[200] K. Xu, C.C. Lin, X. Xie, and A. Meijerink, *Chem. Mater.* 29, 4265 (2017).

[201] W. Liu, Q. Lin, H. Li, K. Wu, I. Robel, J.M. Pietryga, and V.I. Klimov, *J. Am. Chem. Soc.* 138, 14954 (2016).

[202] D. Parobek, B.J. Roman, Y. Dong, H. Jin, E. Lee, M. Sheldon, and D.H. Son, *Nano Lett.* 16, 7376 (2016).

[203] K. Xu, C.C. Lin, X. Xie, and A. Meijerink, *Chem. Mater. Pub. Am. Chem. Soc.* 4265 (2017).

[204] S.D. Adhikari, S K. Dutta, A. Dutta, A.K. Guria, and N. Pradhan, *Angew. Chemie Int. Ed.* (2017).

[205] D. Parobek, Y. Dong, T. Qiao, and D.H. Son, *Chem. Mater* 30, 2939 (2018).

[206] S. Zou, Y. Liu, J. Li, C. Liu, R. Feng, F. Jiang, Y. Li, J. Song, H. Zeng, M. Hong, and X. Chen, *J. Am. Chem. Soc.* 139, 11443 (2017).

[207] Q.A. Akkerman, D. Meggiolaro, Z. Dang, F. De Angelis, and L. Manna, *ACS Energy Lett.* 2, 2183 (2017).

[208] J.S. Yao, J. Ge, B.N. Han, K.H. Wang, H.B. Yao, H.L. Yu, J.H. Li, B.S. Zhu, J.Z. Song, C. Chen, Q. Zhang, H.B. Zeng, Y. Luo, and S.H. Yu, *J. Am. Chem. Soc.* 140, 3626 (2018).

[209] R. Begum, M.R. Parida, A.L. Abdelhady, B. Murali, N.M. Alyami, G.H. Ahmed, M.N. Hedhili, O.M. Bakr, and O.F. Mohammed, *J. Am. Chem. Soc.* 139, 731 (2017).

[210] T.J. Milstein, D.M. Kroupa, and D.R. Gamelin, *Nano Lett.* 18, 3792 (2018).

[211] T.C. Jellicoe, J.M. Richter, H.F. Glass, M. Tabachnyk, R. Brady, S.E. Dutton, A. Rao, R.H. Friend, D. Credgington, N.C. Greenham, and M.L. Bohm, *J. Am. Chem. Soc.* 138, 2941 (2016).

[212]   A. Wang, Y. Guo, F. Muhammad, and Z. Deng, *Chem. Mater.* 29, 6493 (2017).

[213]   A.B. Wong, Y. Bekenstein, J. Kang, C.S. Kley, D. Kim, N.A. Gibson, D.D. Zhang, Y. Yu, S.R. Leone, L.W. Wang, A.P. Alivisatos, and P.D. Yang, *Nano Lett.* 18, 2060 (2018).

[214]   D.S. Dolzhnikov, C. Wang, Y. Xu, M.G. Kanatzidis, and E.A. Weiss, *Chem. Mater.* 29, 7901 (2017).

[215]   X. Wu, W. Song, Q. Li, X. Zhao, D. He, and Z. Quan, *Chem. Asian J.* 13, 1654 (2018).

[216]   J. Pal, A. Bhunia, S. Chakraborty, S. Manna, S. Das, A. Dewan, S. Datta, and A. Nag, *J. Phys. Chem. C* 122, 10643 (2018).

[217]   B. Pradhan, G.S. Kumar, S. Sain, A. Dalui, U.K. Ghorai, S.K. Pradhan, and S. Acharya, *Chem. Mater.* 30, 2135 (2018).

[218]   J. Pal, S. Manna, A. Mondal, S. Das, K.V. Adarsh, and A. Nag, *Angew. Chemie-Int. Ed.* 56, 14187 (2017).

[219]   L. Zhou, Y.-F. Xu, B.-X. Chen, D.-B. Kuang, and C.-Y. Su, *Small* 14, 1703762 (2018).

[220]   Y. Bekenstein, J.C. Dahl, J. Huang, W.T. Osowiecki, J.K. Swabeck, E.M. Chan, P. Yang, and A.P. Alivisatos, *Nano Lett.* 18, 3502 (2018).

[221]   Y. Hassan, Y. Song, R.D. Pensack, A.I. Abdelrahman, Y. Kobayashi, M.A. Winnik, and G.D. Scholes, *Adv. Mater.* 28, 566 (2016).

[222]   J. Shamsi, Z.Y. Dang, P. Ijaz, A.L. Abdelhady, G. Bertoni, I. Moreels, and L. Manna, *Chem. Mater.* 30, 79 (2018).

Chapter 3

Properties of Metal Halide Perovskites

3.1 Structural Properties and Emerging Metal Halide Perovskites

3.1.1 *Crystal Structure and Characterization Methods*

The crystal structure of crystalline solids is most often determined using X-ray diffraction (XRD). By examining the diffraction pattern based on Bragg's law, one can identify the crystal phase of the material. Bragg's law describes the condition on θ, the glancing angle or Bragg angle, for the constructive interference to be at its strongest, as follows:

$$2d \, sin \, \theta = n \, \lambda \tag{3.1}$$

where n is a positive integer, λ is the wavelength of the incident X-ray wave, and d is the interplanar distance between lattice planes. An XRD pattern is obtained by measuring the intensity of scattered X-ray waves as a function of scattering angle (θ). Very strong intensities, known as Bragg peaks, are observed in the diffraction pattern at the points where the scattering angles satisfy the Bragg condition.

XRD can be used to study both bulk crystals and nanocrystals (NCs). For NCs or nanoparticles such as quantum dots (QDs), small angle scattering is sensitive to the average interparticle distance while wide-angle diffraction is sensitive to the atomic structure of the NCs. The widths of the diffraction lines contain information about the size, size distribution, defects, and strain in the NCs. As the size of the NCs decreases, the line width is broadened due to loss of long-range order relative to the bulk. The XRD line width can be used to estimate the size of the NCs via the Debye–Scherrer formula:

$$D = 0.9 \ \lambda/(\beta \cos \theta) \tag{3.2}$$

where D is the NC diameter, λ is the wavelength of light, β is the full width at half maximum (FWHM) of the peak in radians.

Among the metal halide perovskites (MHPs), methylammonium (MA) lead iodide, $CH_3NH_3PbI_3$, or $MAPbI_3$, has been the most investigated. Figure 3.1 shows a photograph of a large single crystal and representative XRD patterns of $MAPbI_3$, along with an illustration of the structure of (100) facet-oriented $MAPbI_3$ crystal and UV-visible absorption spectrum [1]. X-ray single crystal structural analyses reveal that the crystal belongs to the space group of I4/mcm (140) and three-dimensional Pb-I links form the integral structural framework, and the MA^+ ion is in a cage made by four PbI_6 octahedra. The UV-visible absorption spectrum shows an optical bandgap of 836 nm or 1.48 eV, which is the same as the previously reported value [2].

Figure 3.2 shows the unit cells of different phases of $MAPbI_3$ [3]. The crystal phases of $MAPbX_3$ are temperature dependent, going from cubic at high temperature to tetragonal and then orthorhombic with decreasing temperature. For the cubic phase at high temperature, the $CH_3NH_3^+$ ion is polar with C_{3v} symmetry, which should result in disordered cubic phase [4]. In addition, the halogen ions are also disordered in the cubic phase, as shown in Figure 3.2(a). With decreasing temperature, the cubic phase is transformed into the tetragonal phase, as shown in Figure 3.2(b), and the I^- ions are ordered. As the temperature decreases further, the tetragonal phase is transformed into the orthorhombic phase, in which the $CH_3NH_3^+$ ions are ordered as shown in Figure 3.2(c).

Neutron powder diffraction is another useful tool for structural determination and has been applied to study structure and phase transitions in MHPs. For example, it was used to reveal rotational distortion of the PbI_6 octahedra in $MAPbI_3$ during phase transitions from cubic to tetrahedral

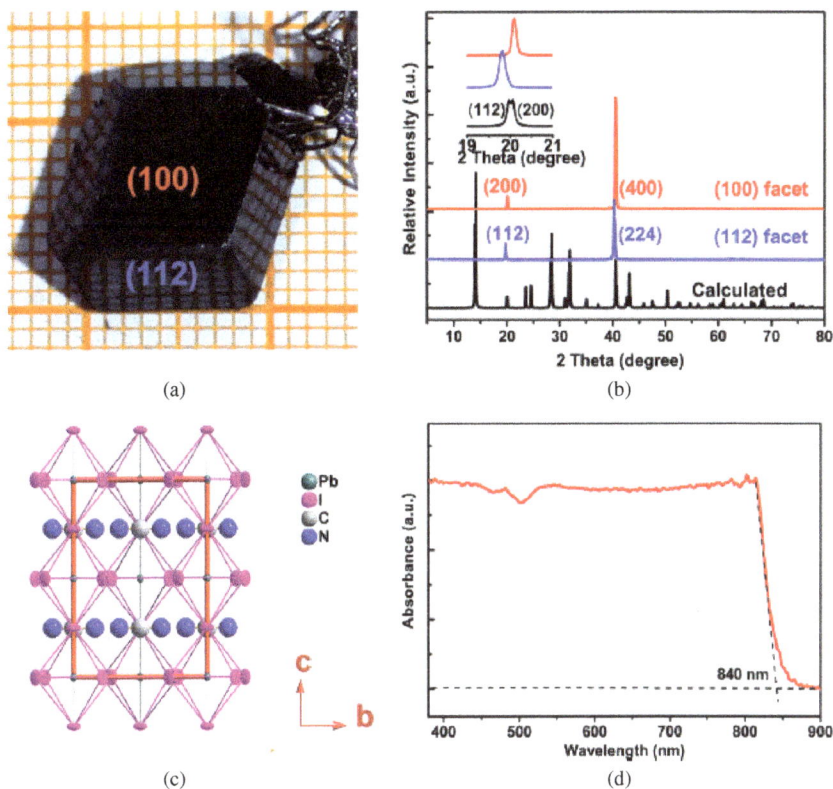

Figure 3.1. Characterization of as-grown MAPbI₃ bulk single crystal. (a) Photograph of a MAPbI₃ bulk single crystal grown through the BSSG method. Natural facets of the MAPbI₃ bulk single crystal were indexed. The unit length of each grid on the coordinate paper is 1 mm. (b) XRD patterns of the rhombic (red) and the parallelogram (blue) natural crystallographic facets of a MAPbI₃ single crystal. The calculated powder XRD pattern (black) is also shown as a reference. The inset shows the details of distinguishing (100) and (112) facets. (c) Drawing of the structure of (100) facet-oriented MAPbI₃ crystal (partial *N* atoms are made invisible in order to expose the C atoms). (d) UV-visible absorption spectrum for a (100)-cut 1 mm-thick MAPbI₃ single crystal plate. Reproduced from Ref. [1].

and orthorhombic, as shown in Figure 3.3 [5]. The study was assisted with isotope substitution where partial or complete replacement of H by D was done for the MA cations. In the cubic *Pm-3m* structure, the octahedra are not rotated. In the I4/mcm structure, the octahedra in neighboring planes along the c axis rotate in the opposite directions, whereas in the *P4/mbm*

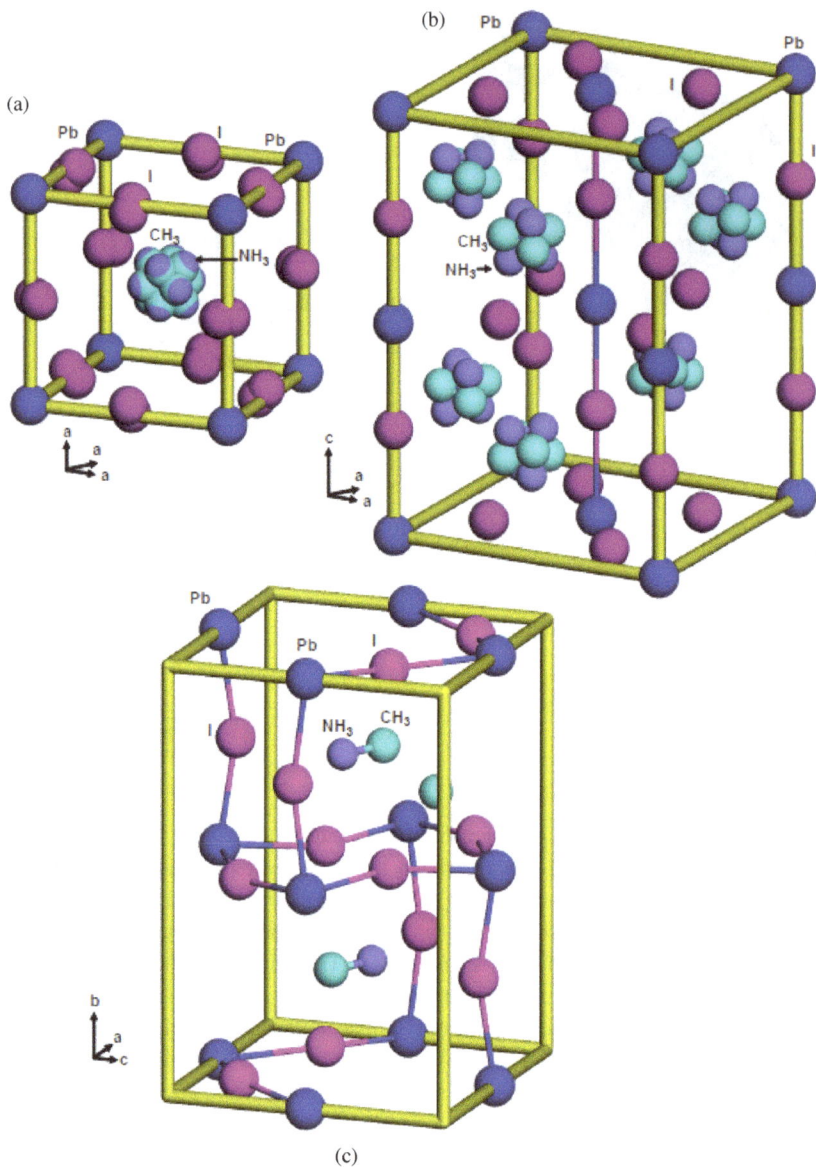

Figure 3.2. Structure models of $CH_3NH_3PbI_3$ with (a) cubic, (b) tetragonal, and (c) orthorhombic structures. Reproduced from Ref. [3].

(a)

(b)

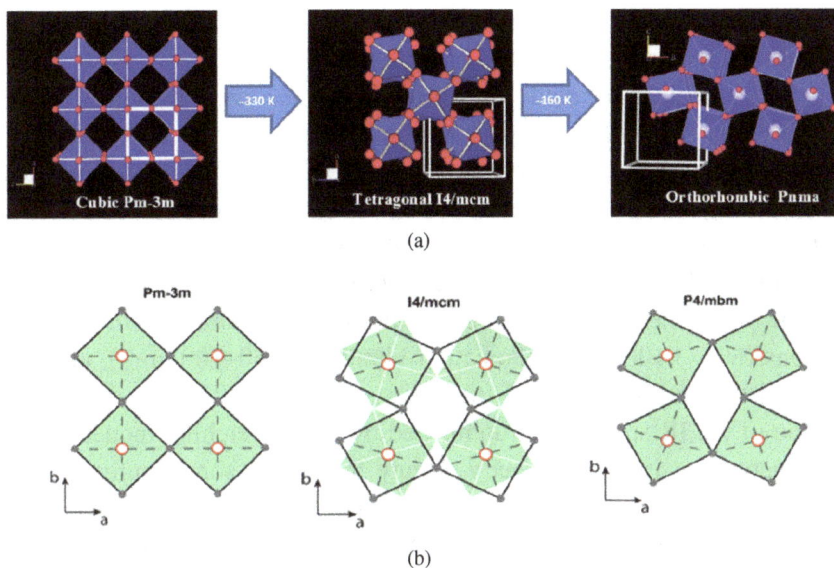

Figure 3.3. The crystal structures adopted by MAPbI$_3$. (a) The PbI$_6$ octahedra are blue and the iodine atoms are red. The cubic to tetragonal transition is due to the R$_4^+$ rotational distortion mode, and the tetragonal to orthorhombic transition is primarily associated with a combination of the R$_4^+$ and M$_3^+$ distortion modes. In this figure, the MA cations are not shown to better highlight the distortions of the Pb-I network due to the structural phase transitions. (b) The relative rotations of neighboring layers of PbI$_6$ octahedra along the c axis are shown as filled green squares and unfilled black squares. In the cubic *Pm-3m* structure the octahedra are not rotated. In the *I4/mcm* structure the octahedra in neighboring planes along the c axis rotate in the opposite sense, whereas in the *P4/mbm* structure the neighboring planes of octahedra along the c axis rotate in the same sense. Reproduced from Ref. [5].

structure, the neighboring planes of octahedra along the c axis rotate in the same direction.

Infrared (IR) absorption and Raman scattering are two common spectroscopic techniques that can provide structural information about molecules as well as solid crystalline materials. In particular, they can afford information about characteristic phonon modes of solids as well as vibrational modes of molecules on the surface or intercalated inside solids, which are molecular specific or selective. Since the spectroscopic selection rules between IR and Raman spectroscopy are often complementary to each other, they can provide information about different phonon or vibrational modes.

For example, a detailed study of the vibrational properties of MAPbI$_3$ has been carried out in the 6–3,500 cm^{-1} range from a combination of theory and experiment based on factor group analysis, first-principles calculations, and low-temperature infrared spectra [6]. Good agreement between experiment and theory led to the assignment of most of the features in the IR spectrum to specific vibrational modes. The IR spectrum can be partitioned into three regions: the internal vibrations of the MA cations (800–3,100 cm^{-1}), the cation librations (140–180 cm^{-1}), and the internal vibrations of the PbI$_3$ network (<100 cm^{-1}). The low-frequency region of the IR spectrum is dominated by Pb–I stretching modes of the PbI$_3$ network with B_u symmetry. Figure 3.4 shows some representative normal modes of vibrations of the PbI$_3$ network in MAPbI$_3$ [6].

In Raman scattering measurement, a single frequency light, usually from a single mode laser source, illuminates the sample and inelastically scattered light is measured off angle with respect to the incident light to minimize Rayleigh scattering. The scattered light with lower (Stokes scattering) or higher (anti-Stokes scattering) frequencies is measured with a photodetector. The energy difference between the scattered ($h\nu_s$) and incident light ($h\nu_i$), so-called *Raman shift* (usually given in wave number cm^{-1} = 1/λ with wavelength λ expressed in cm) equals to the vibrational or phonon frequencies of the sample ($h\nu_v$), as long as selection rules allow. For Stokes Raman scattering:

$$h\nu_v = h\nu_{ph} - h\nu_s \qquad (3.3)$$

The spectrum is usually expressed in terms of intensity of the Raman scattered light as a function of Raman shift ($h\nu_v$).

The low-frequency resonant Raman spectrum of MAPbI$_3$ has been studied both experimentally and computationally [7]. The experimentally measured Raman spectrum was assigned with the assistance of DFT simulations of suitable periodic and model systems, as shown in Figure 3.5. The bands at 62 and 94 cm^{-1} are assigned, respectively, to the bending and stretching of the Pb–I bonds, and are considered as diagnostic modes of the inorganic cage. Vibrational frequencies in the 119 and 154 cm^{-1} range are

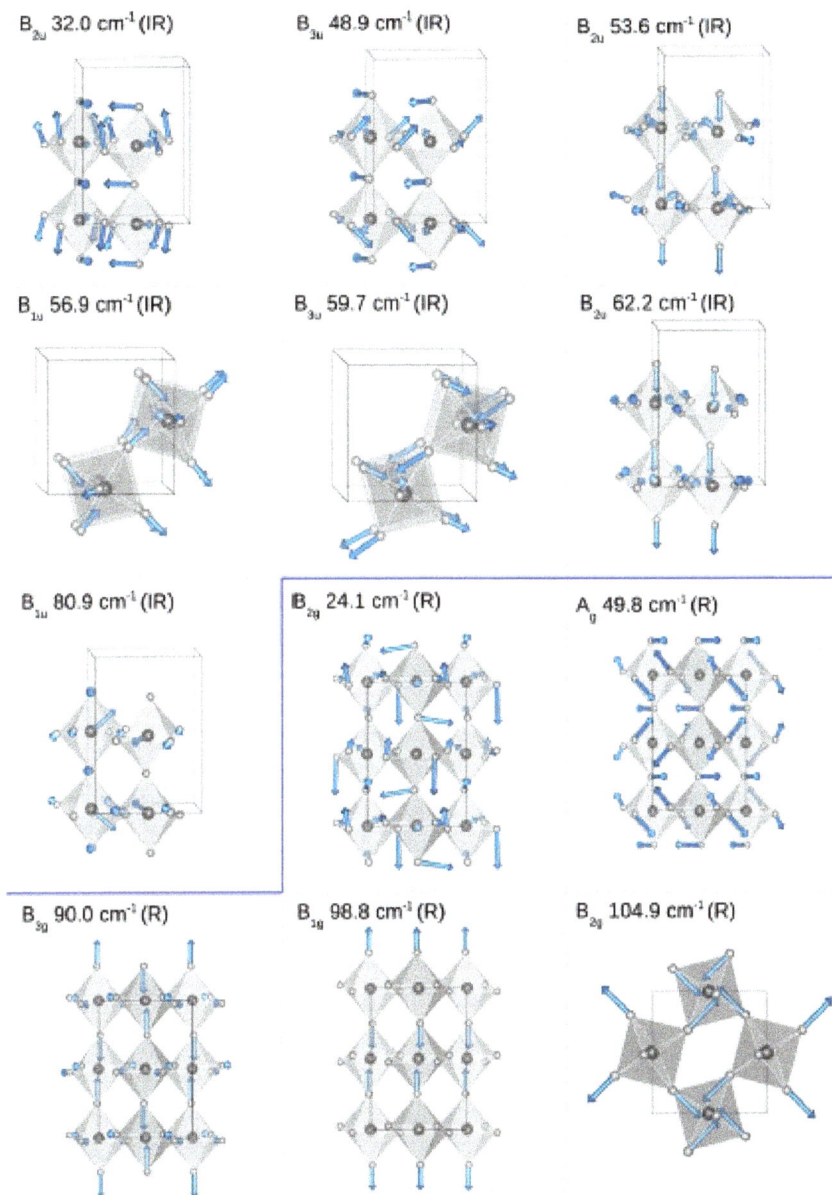

Figure 3.4. Schematic representation of some representative normal modes of vibration of the PbI$_3$ network in MAPbI$_3$. The PbI$_6$ octahedra are shaded, the Pb and I atoms are in black and gray color, respectively, and the MA cations are not shown for clarity. The symmetry, (calculated) frequency, and IR or Raman (R) activity of each mode are also indicated. Reproduced from Ref. [6].

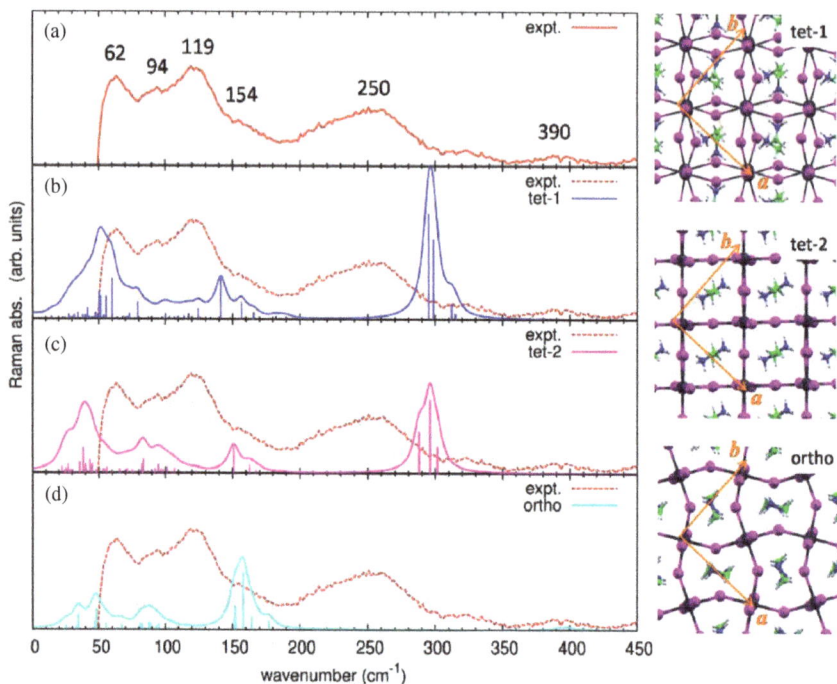

Figure 3.5.   Left: Experimental resonant (a) and non-resonant calculated (*tet-1, tet-2, ortho*) Raman spectra of MAPbI$_3$ (b–d). The two tetrahedral (*tet-1 and tet-2*) and orthorhombic (*ortho*) optimized structures parallel to the ab plane are shown on the right panel. The periodic structures and Raman spectra were calculated by the Quantum Espresso program package. Reproduced from ACs from Ref. [7].

assigned to librations of the organic cations (MA). The broad, unstructured 200–400 cm^{-1} features are assigned to the torsional mode of the MA, which are proposed as markers of the orientational disorder of the material [7].

Raman spectroscopy is also convenient for *in situ* studies. For example, *in situ* Raman has been used to study the thermal stability of CsPbX$_3$ NCs between the cryogenic temperature and high temperature [8]. The low-frequency Raman signatures of CsPbX$_3$ NCs dramatically evolve and provide information about their crystal structures and phase transitions. For instance, as shown in Figure 3.6, the NCs undergo a state of high degree of disorder with featureless Raman spectra before being thermally decomposed.

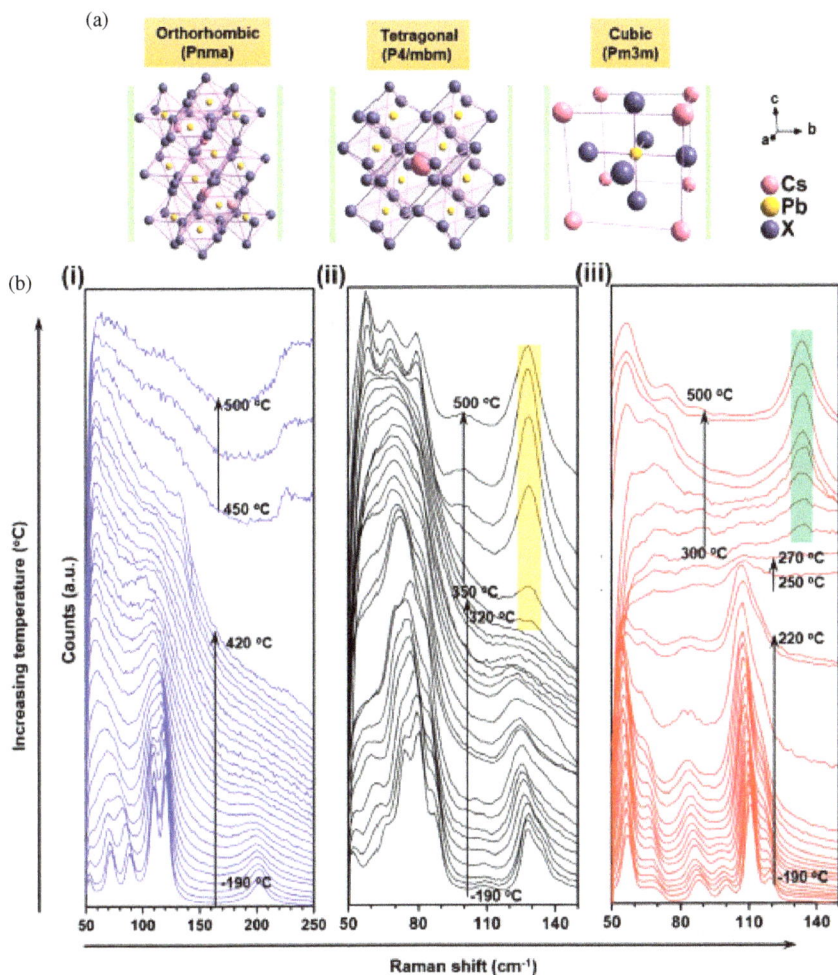

Figure 3.6. Temperature-dependent Raman profiles of $CsPbX_3$ NCs. (a) Schematic illustration of structures of orthorhombic, tetragonal, and cubic $CsPbX_3$ NCs. (b) Temperature-dependent Raman spectra at representative temperatures of (i) $CsPbCl_3$, (ii) $CsPbBr_3$, and (iii) $CsPbI_3$ NCs. *In situ* Raman measurements were started from 25°C, and then the temperature was reduced to −190°C to monitor the spectral evolution of $CsPbX_3$ NCs in low-temperature regions. The marked yellow and green regions in (ii) and (iii) represent the decomposition of $CsPbBr_3$ and $CsPbI_3$ NCs. Reproduced from ACS from Ref. [8].

Besides providing structural information about the MHPs, IR and Raman can also be used to gain information about molecules, such as passivation ligands, on the surface of MHPs and their interactions. For example, both IR and Raman have been used to confirm the presence of passivating ligands, a combination of (3-aminopropyl)triethoxysilane (APTES) and phosphonic acids (PAs), on the surface of MAPbBr$_3$ perovskite quantum dots (PQDs), as shown in Figure 3.7 [9]. In particular, the characteristic bands of P–OH, P=O, Si–O, and O–H in the FTIR spectra demonstrate the presence of the PA and APTES as capping ligands on the PQD surface (samples were measured after washing away free ligands in solution).

3.1.2 *Morphology and Characterization Methods*

Solid materials can exist in various morphologies. This is also true for MHPs. Typical morphologies include bulk single crystals, polycrystals (many in thin films), nanostructures in various sizes and shapes. For example, Figure 3.8 shows some typical SEM images of several MHPs synthesized using a solution method, including $CH_3NH_3SnI_3$, $HC(NH_2)_2SnI_3$, $CH_3NH_3PbI_3$, and $HC(NH_2)_2PbI_3$ that reveal different crystal habits [10]. The crystal size can vary significantly from sample to sample and depends strongly on preparation methods used. As will be discussed later (Section 3.6), NCs or QDs can also be prepared and exhibit properties different from bulk crystals.

When examined closely via microscopy or spectroscopy, one often observes heterogeneity on multiple length scales. For example, Figure 3.9 shows electron microscopy images of MHPs that reveal disorder at the grain and sub-grain levels [11]. Gains containing sub-grain features such as crystallites and twin domains may be related to local strain fields and can alter the diffusion pathways of electrons and ions. The interfaces between grains and their overall 3D distribution in the microstructure of the sample influence non-radiative recombination, ion migration, infiltration of species such as oxygen and moisture, and device operation. The morphology strongly depends on the methods of sample preparation.

Morphological information can also be obtained from optical or spectroscopic imaging such as confocal PL microscopy. For example, Figure 3.10 shows a collection of different PL images, PL lifetime images,

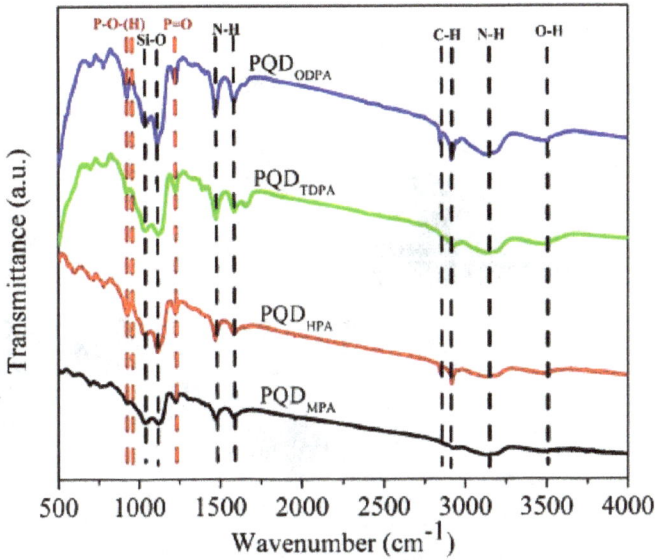

(a)

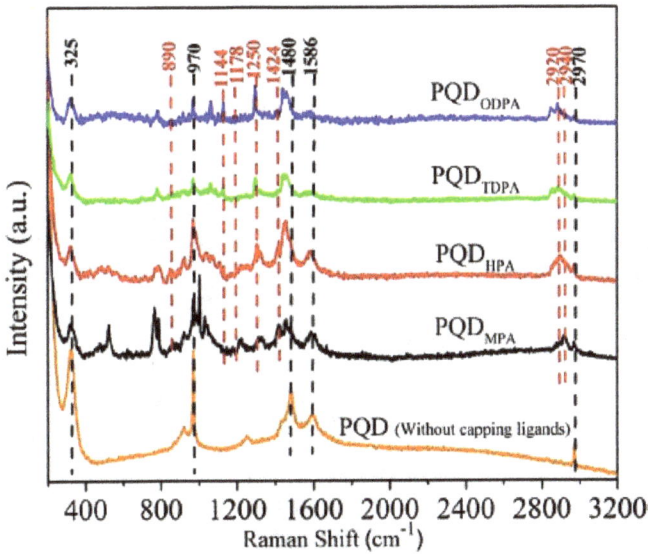

(b)

Figure 3.7.   IR and Raman spectrum of MXPbBr$_3$ PQDs passivated with APTES and PAs with different alkane chain lengths, including methylphosphonic acid (MPA), *n*-hexylphosphonic acid (HPA), 1-tetradecylphosphonic acid (TDPA), and *n*-octadecylphosphonic acid (ODPA). Reproduced from Ref. [9].

Figure 3.8. Typical SEM images showing the crystal habits of several MHPs obtained from a solution method: (a) $CH_3NH_3SnI_3$, (b) $HC(NH_2)_2SnI_3$, (c) $CH_3NH_3PbI_3$, (d) $HC(NH_2)_2PbI_3$. Reproduced from ACS from Ref. [10].

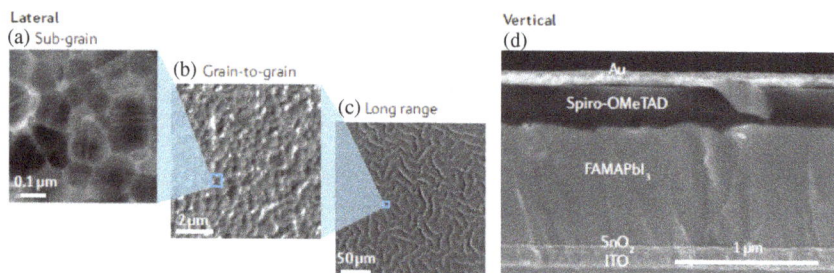

Figure 3.9. The hierarchy of heterogeneity in halide perovskites. (a–c) The electron microscopy images exemplify the sub-grain (panel a), grain-to-grain (panel b), and long-range (panel c) disorder within halide perovskites. (d) A cross-sectional scanning electron microscopy image of a complete, high-performance perovskite solar cell. Vertical heterogeneity in the perovskite grain morphology is evident despite the device having an excellent power conversion efficiency of 23.32%. FA$^+$, formamidinium; ITO, indium-doped tin oxide; MA$^+$, methylammonium; spiro-OMeTAD, 2,2′,7,7′-tetrakis (N,N-di-pmethoxyphenylamine)-9,9′-tetrakis(*N,N*-di-*p*-methoxyphenylamine)-9,9′-spirobifluorene. Reproduced from Ref. [11].

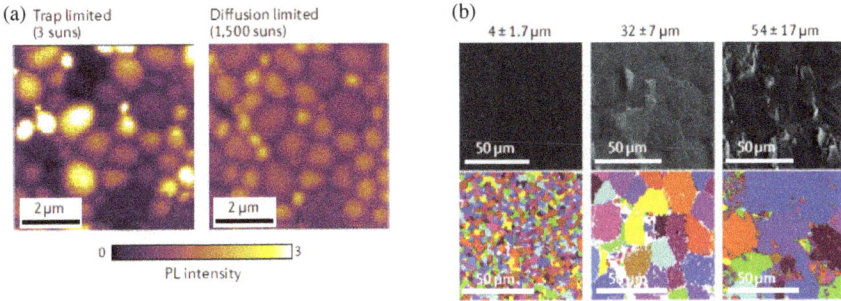

Figure 3.10. Grain-to-grain heterogeneity. (a) Photoluminescence (PL) maps of MAPbI$_3$ (where MA$^+$ is methylammonium) films at different photon fluxes. The incident excitation intensity influences the transport mechanisms of the charge carriers: at low intensity (left), the charge carriers are trap limited, whereas at higher intensities (right), the charge carriers are diffusion limited. (b) Scanning electron microscopy (SEM) images (top) and electron backscattering diffraction (EBSD) plots (bottom) of MAPbBr$_3$ perovskite thin-film samples with different grain sizes (stated at the top of the corresponding SEM image). The heterogeneity in the crystallographic orientation increases with grain size, and amorphous grain boundaries (white regions) are present in the sample with 32-μm-sized grains. Note that the colors in the EBSD plots do not represent a specific crystallographic orientation. (c) The plot of yield–mobility product as a function of grain size in a MAPbI$_3$ perovskite shows that the mobility decreases as the average grain size decreases. The symbols represent the experimental data (with error bars indicating the standard deviation in both the grain size and yield–mobility product). The dashed line is the best fit of the Kubo relationship for the first two data points, whereas the solid line is the best fit for all the data. (d) Single-photon (left) and two-photon (right) PL maps that probe the surface and bulk radiative recombination emission, respectively, in a MAPbI$_3$ thin film. (e) Spectrally integrated cathodoluminescence (CL) maps of a CsPbIBr$_2$ film acquired in an electron microscope. The top part shows the high-energy spectral distribution, which is probably from a Br$^-$-rich phase, whereas the middle part shows lower-energy emission and is likely from an I$^-$-rich phase. The bottom map is a superposition of the two CL images (green indicates high-energy emission and red indicates lower-energy emission) and shows that the I$^-$-rich regions are pushed toward the grain boundaries. (f) A PL intensity map (left) of a MAPbI$_3$ thin film overlaid with a fluorescence lifetime imaging microscopy map. The brightness indicates the PL emission intensity, whereas the color indicates the lifetime. Three grains are emphasized with respect to their neighbors by line profiles L1, L2, and L3; each of these three grains has a low PL intensity compared with its neighboring grains (top right part). However, as indicated by the line traces on the bottom right, the central grains in L1, L2, and L3 exhibit sustained, longer, or shorter lifetimes, respectively, than the neighboring grains. Reproduced from Ref. [11].

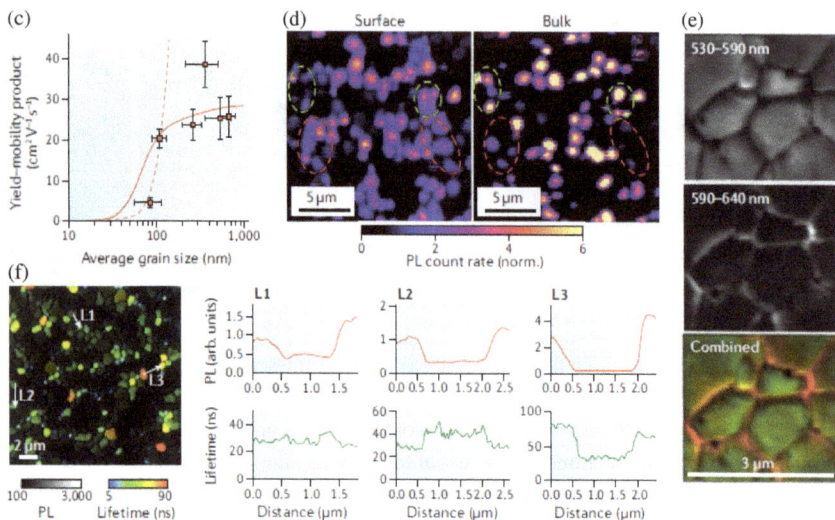

Figure 3.10.   (*Continued*)

and SEM images to show heterogeneity, especially grain to grain, for different MHP films on different length scales [11]. Figure 3.10(a) shows the PL heterogeneity is dependent on the light intensity used and some perovskite grains only emit efficiently under high-fluence conditions. This was taken as an indication that trap-mediated pathways become saturated at higher excitation densities and thus the PL response is dominated by charge-carrier diffusion. Therefore, at sufficiently low excitation densities, such as normal solar light, the PL distribution is typically dominated by trap-assisted recombination. Figure 3.10(d) shows a comparison between two-photon and one-photon PL maps, where some grains appear brighter at the surface than in the bulk (indicated by the red circles), while other grain clusters demonstrate the opposite phenomenon (indicated by the green circles). Such depth-dependent PL results reveal buried recombination processes and suggest that there is both lateral and vertical heterogeneity in the radiative recombination. Figure 3.10(f) shows confocal fluorescence-lifetime imaging microscopy (FLIM) maps of a $MAPbI_3$ perovskite thin film that reveal a wide grain-to-grain distribution of both the PL intensity and lifetime.

3.1.3 *Metal Halide Double Perovskites*

Metal halide double perovskites (MHDPs) with a general formula of $A_nB'B''X_m$, where B' and B" represent two different metal cations, n and m are integers, have attracted increasing attention lately as a promising and stable substitution for Pb-based halide perovskites. In DMHPs, B and B' are monovalent and trivalent metal cations, such as Ag^+, Cu^+, Bi^{3+}, Sb^{3+}, and In^{3+} that can be accommodated in the lattice. Their oxide counterparts ($A_2BB'O_6$) have been studied extensively in the fields of ferromagnetics, ferroelectrics, colossal magnetoresistance, and multiferroics, with applications in scintillators in radiation detectors. In MHDPs, hetero-substitution of Pb^{2+} ions by other metal ions can maintain similar structural integrity while affording flexibility for varying chemical compositions to alter the properties and functionalities of the materials. Such MHDPs exemplified by $Cs_2AgBiBr_6$ and $(AMP)_4(AgBiI_8) \cdot H_2O$ (AMP for 4-aminopiperidinium) are highly promising materials not only for high efficiency PV devices and (LEDs but also for spintronic and spin-optotronic applications such as spin field-effect transistors (FET) [12,13] and spin-LEDs [14–17]. Figure 3.11 shows the crystal structures of three different MHDPs as compared to the normal simple MHPs [18].

For the $A_2B^+B^{3+}X_6$ 3D halide perovskites, the B^+ and B^{3+} cations exhibit alternating arrangement in the octahedral cavity and have six-fold coordination with the halide ions, respectively. The A site cations are located in cavities formed by octahedrons. If two Pb^{2+} cations are substituted via a vacancy and a quadrivalent cation, such as Sn^{4+}, Ge^{4+}, and Ti^{4+}, $A_2B^{4+}X_6$ can be formed, which is termed "vacancy-ordered double perovskites", and has crystal structure similar to that of the double perovskites. The key difference is that the vacancy-ordered double perovskites have 50% periodic vacancies on the octahedral cavity. On the other hand, if four Pb^{2+} ions are replaced by two vacancies, one divalent cation and two trivalent cations, layered MHDPs with defective 3D structure can be formed, with the general chemical formula of $A_4B^{2+}B^{3+}_2X_{12}$, where B^{2+} could be Sn^{2+}, Ge^{2+}, and Cu^{2+} while B^{3+} could be Sb^{3+}, In^{3+}, and Bi^{3+}. The layered MHDPs are a unique hybrid metal <111>-oriented layered perovskites, which contain a $B^{2+}X_6$ octahedra

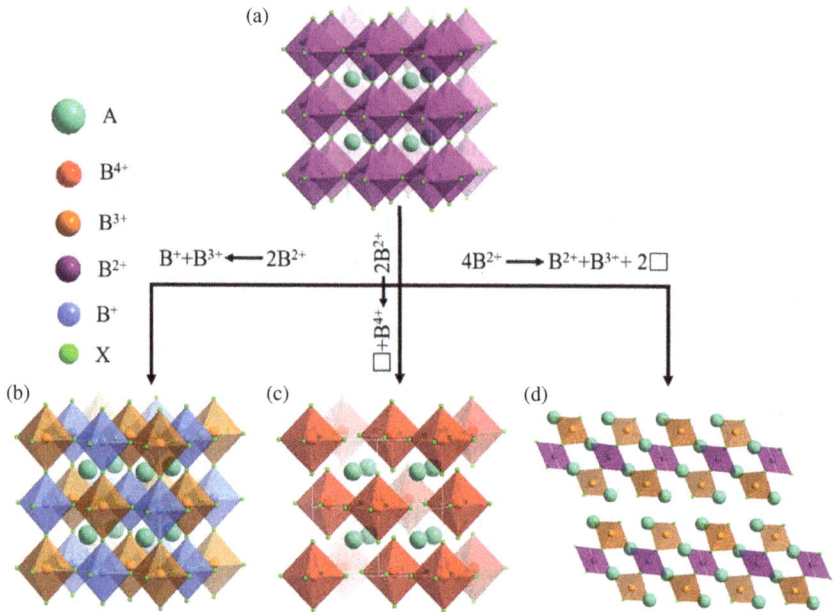

Figure 3.11. The crystal structures of (a) simple halide perovskites ($AB^{2+}X_3$), (b) halide double perovskites ($A_2B^+B^{3+}X_6$), (c) vacancy-ordered halide double perovskites ($A_2B^{4+}X_6$), and (d) layered halide double perovskites ($A_4B^{2+}B^{3+}_2X_{12}$). Reproduced from Ref. [18].

inserted between two layers of $B^{3+}X_6$ octahedrons. Basically, the vacancies generated by B^{2+} site cations result in a collapse of the 3D network into a 2D network [18].

As an example of MHDPs, $Cs_2AgBiBr_6$ has been found to exhibit high stability, long carrier recombination lifetimes, low carrier effective masses, and promising PV performance in solar cell application [19]. It also shows potential for applications in sensors, photodetectors, radiation detectors, and photocatalysis. Doping and alloying have been used to tune the properties of MHDPs. For example, by replacing a fraction of Cs^+ ions with Na^+ or K^+ ions and/or replacing some In^{3+} ions with Bi^{3+} ions, the optical properties of $Cs_2AgInCl_6$ can be altered, including enhanced absorption and PL [20]. The original Cs_2AgInX_3 has no sharp excitonic absorption band and is not photoluminescent due to the indirect bandgap nature of the first excitonic transition. Upon doping or alloying with metal ions such as Bi^{3+}, the electronic structure changed more toward direct bandgap in nature, which results in a strong excitonic absorption around 360 nm and PL around 600 nm attributed to emission from self-trapped exciton.

3.1.4 *Two-dimensional (2D) Metal Halide Perovskites*

Recently, there has been growing interest in two-dimensional (2D) MHPs. Compared to three-dimensional (3D) perovskites, 2D perovskites have similarities and differences [21,22]. In contrast to conventional 3D lead halide perovskites with the chemical formula of ABX_3 where organic or inorganic cations (A) fill the voids between lead halide octahedra $(BX_6)^{4-}$, 2D perovskites use additional larger organic cations (L) to isolate the inorganic metal halide octahedra layers, forming essentially quantum well superlattices. Figure 3.12 shows three representative 2D perovskite structures [21].

If the spacers are hydrophobic, they can help to increase stability of the perovskites toward moisture. Examples of common spacers are given in Figure 3.13 [22].

Furthermore, the extra spacing cations and the resulting asymmetric lattice structures provide additional degrees of freedom to tune the

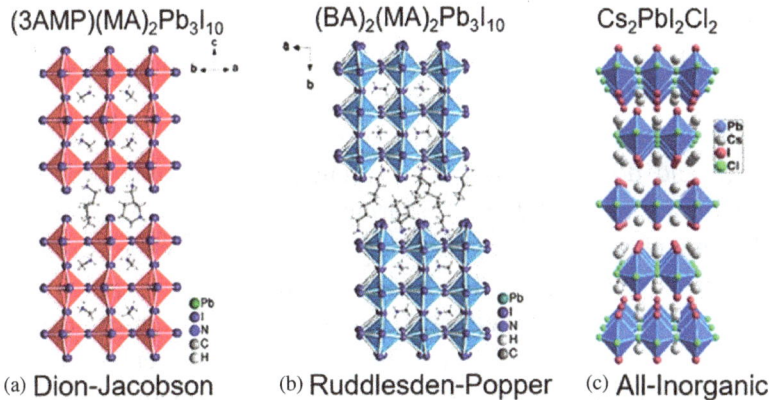

$(3AMP)(MA)_2Pb_3I_{10}$ $(BA)_2(MA)_2Pb_3I_{10}$ $Cs_2PbI_2Cl_2$

(a) Dion-Jacobson   (b) Ruddlesden-Popper   (c) All-Inorganic

Figure 3.12. Crystal structural of typical (a) Dion–Jacobson, (b) Ruddlesden–Popper, and (c) All-inorganic 2D lead halide perovskites. Reproduced from Ref. [21].

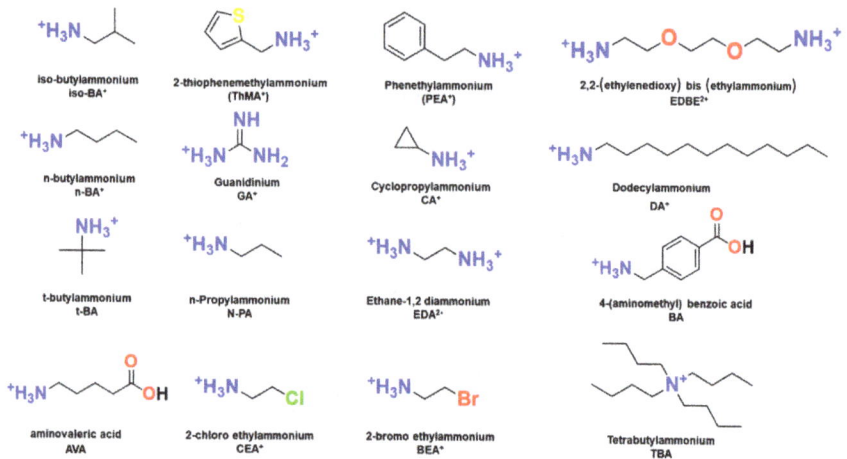

Figure 3.13. Examples of reported organic cations used to construct 2D/3D perovskites. Reproduced from Ref. [22].

properties of the perovskites, including bandgap, exciton binding energy, exciton dynamics, charge carrier transport, and electron–phonon coupling. These characteristics can thus be controlled for desired applications including PV solar cells and LEDs. For example, propargylarnine cation (PPA$^+$) has been employed as an organic spacer to enhance charge carrier transport of quasi-2D (PPA)$_2$(CH$_3$NH$_3$)$_2$Pb$_3$I$_{10}$ thin films and when processed with Pb(SCN)$_2$ as an additive, a solar cell with >15% power conversion efficiency (PCE) was achieved [23]. As shown in Figure 3.14, Pb(SCN)$_2$ has a strong influence on the film morphology and charge transport as well as device performance.

It has been of strong interest to combine the advantages of 2D perovskites and double perovskites to generate 2D double perovskites, especially without lead. Besides sharing several key advantages with 3D MHPs, including unique spin-dependent properties such as Rashba and Dresselhaus splitting of degenerate electronic states [24–30], 2D MHDPs have demonstrated high stability, efficient in-plane charge transport, possibility to switch between direct and indirect bandgap transitions, and additional flexibility for manipulating structure and composition to tune spin-optotronic properties [14,16,31–33]. For example, several different metal ions can be used for the B' or B" positions, and

Figure 3.14. SEM images of (a) pristine quasi-2D $(PPA)_2(MA)_2Pb_3I_{10}$ thin film and quasi-2D $(PPA)_2(MA)_2Pb_3I_{10}$ thin films processed with (b) 1% $Pb(SCN)_2$, (c) 3% $Pb(SCN)_2$, and (d) 5% $Pb(SCN)_2$ additives, respectively. Reproduced from Ref. [23].

many organic cations can be used as the A component, such as 1,4-cyclohexanediammonium [34], AMP [17], 3-iodo-propan-1-ammonium [35], β-methylphenylethylammonium (β-MPEA) [36]. If necessary, one can also introduce other metal cations as dopants to replace B' or B" to further alter the properties of the MHDPs [37]. Most studies to date have focused on 2D MHDPs with the A component being an organic cation, that is, hybrid MHDPs. Compared to the limited choices of inorganic A components, such as Cs^+, many organic cations can serve as potential A components, affording significant flexibility to vary the structure and properties of the resulting 2D hybrid MHDPs. Figure 3.15 shows examples of 2D vs 3D double perovskites [16].

3.1.5 *Lead-free Metal Halide Perovskites*

A major concern of lead-based MHPs is the high toxicity of Pb that is undesirable for applications. As a result, there is significant interest in

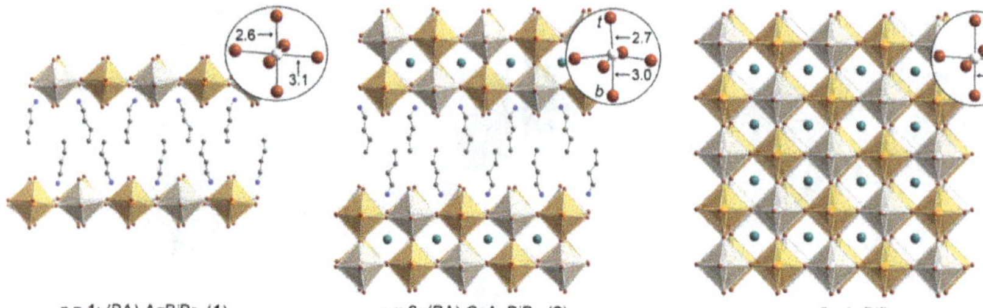

$n = 1$; $(BA)_4AgBiBr_8$ (1)	$n = 2$; $(BA)_2CsAgBiBr_7$ (2)	$n = \infty$; $Cs_2AgBiBr_6$
(a)	(b)	(c)

Figure 3.15. Single-crystal X-ray structures (298 K) of the (001) layered double perovskites (a) $(BA)_4AgBiBr_8$ (**1**; $BA = CH_3(CH_2)_3NH_3^+$) and (b) $(BA)_2CsAgBiBr_7$ (**2**), and the 3D double perovskite (c) $Cs_2AgBiBr_6$. Insets show the Ag coordination sphere with select bond distances in Ångstroms. In the inset for panel B, t denotes a terminal bromide and b denotes a bridging bromide. Orange, white, turquoise, brown, blue, and gray spheres represent Bi, Ag, Cs, Br, N, and C atoms, respectively. H and disordered atoms omitted for clarity. Reproduced with permission from Ref. [16].

developing MHPs with Pb replaced by less toxic metals such as Sn or Bi. The PV efficiency is generally lower for non-lead based MHPs. For example, Figure 3.16 lists several potential or demonstrated organic and inorganic cations for the A site and the bandgap of a number of MHPs with low-toxicity constituents, such as Sn/Ge-based halides, some double perovskites, and some Bi/Sb-based halides [38]. Many of these have direct bandgaps of around 1.1 to 2.0 eV and are potential solar absorbers for PV applications.

Figure 3.17 shows a comparison of the structure, efficiency, and stability of different lead-free MHPs and devices [38]. The crystal structures of lead-free MHPs can be the same as or different from that of lead-based MHPs, depending on the particular A and X components involved. Examples of crystal structures are shown for $FASnI_3$, $Cs_2AgBi(ClBr)_6$, and $Cs_3Bi_2I_9$ [38]. The PCE for lead-free MHPs is usually much less than 50% of the lead-based MHPs. The device stability depends strongly on the specific composition of the MHPs, with some lead-free MHPs showing excellent stability. Except for PV devices, many different lead-free MHPs have also been explored for other applications as summarized in Chapter 4.

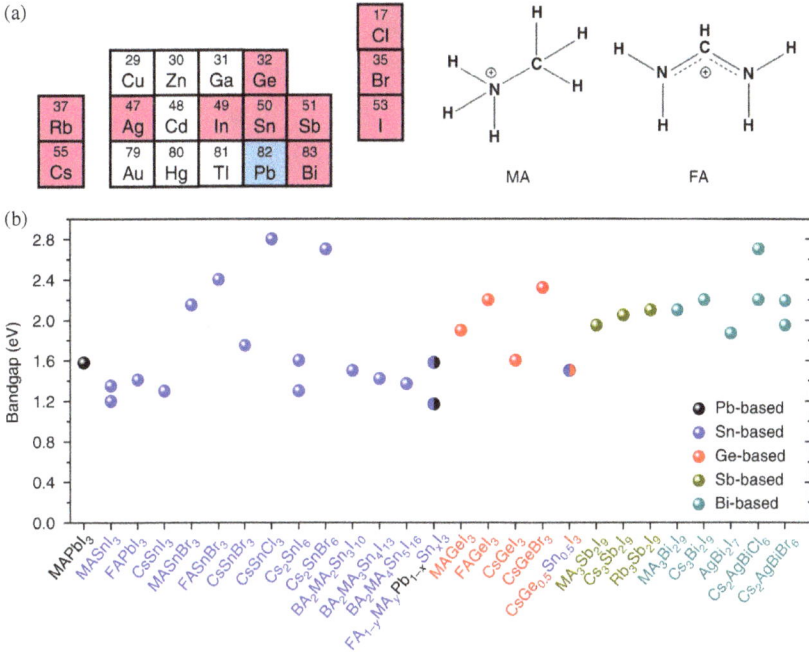

Figure 3.16. Potential materials as solar cell absorbers. (a) Potential A-site cations (organic MA and FA or inorganic Cs and Rb), metals, and halides (I, Br, Cl) for perovskite structure. (b) Bandgaps of various materials. Reproduced from Ref. [38].

3.2 Electronic Properties

3.2.1 *Electronic Structures*

One of the most powerful and commonly used techniques to study the electronic structure of solid materials is X-ray photoelectron spectroscopy (XPS). XPS is based on measurement of the kinetic energy of photoelectrons generated when the sample is illuminated with soft (1.5 kV) X-ray radiation in an ultrahigh vacuum (UHV) [39]. If one X-ray photon with energy $h\nu$ is used to excite an atom in its initial state with energy E_i and to eject an electron with kinetic energy, E_{kin}, with the atom ending in a final state with energy E_f, one would have the following equation based on total energy conservation:

$$H\nu + E_i = E_{kin} + E_f \tag{3.4}$$

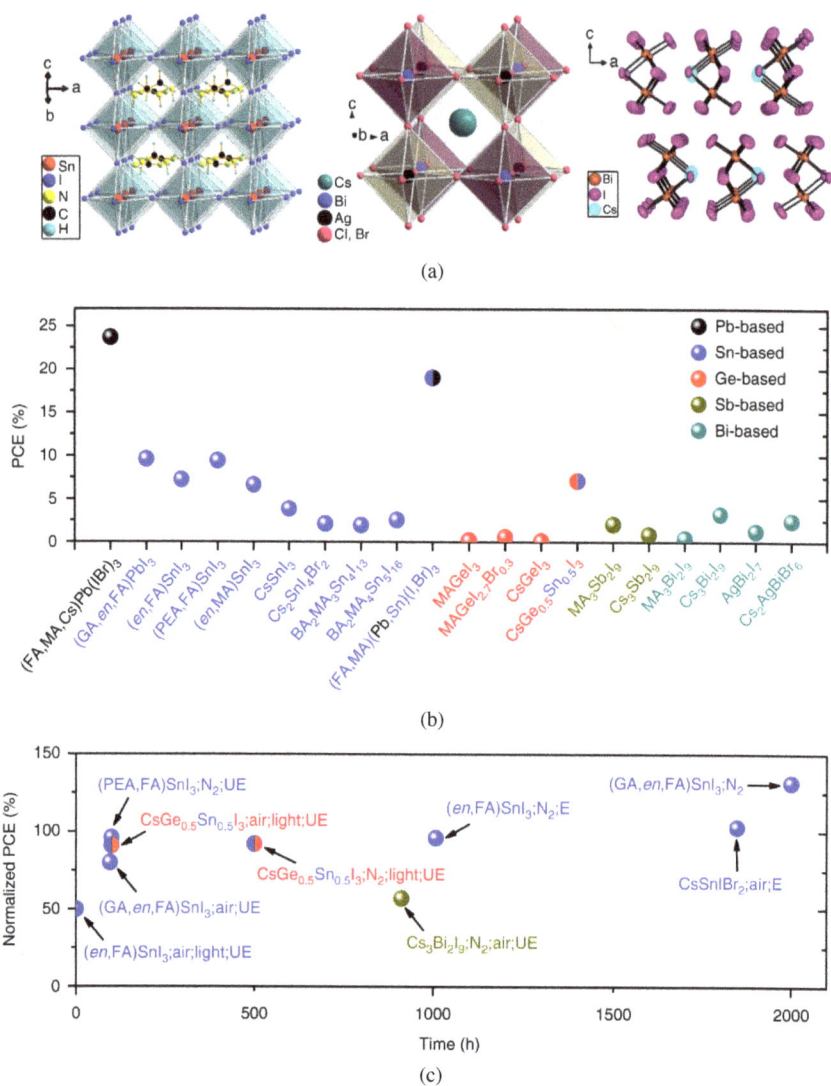

Figure 3.17. Comparison of structure, efficiency, and stability of different materials and devices. (a) Crystals structures of $FASnI_3$ (left), $Cs_2AgBi(ClBr)_6$ (middle), and $Cs_3Bi_2I_9$ (right). (b) Record efficiencies of representative solar cells using Pb, Sn, Ge, Sb, and Bi-based absorbers. (c) Stability of representative Sn, Ge, and Bi-based perovskite solar cells under different conditions, i.e., encapsulated (E), unencapsulated (UE), ambient air, glovebox (N_2), and light soaking. Reproduced from Ref. [38].

The difference between the photon energy and the electron kinetic energy is called the *binding energy* of the orbital from which the electron is ejected, which, based on Eq. (3.4), is equal to $E_f - E_i$. Since the photon energy of the X-ray is known and the electron kinetic energy can be measured, the binding energy can be determined and gives the energy difference between the final and initial states of the atom involved in the transition. This binding energy is characteristic of different orbitals of specific elements and is roughly equal to the Hartree–Fock energy of the electron orbital. Therefore, peaks in the XPS spectrum can be identified with specific atoms and surface composition can be analyzed. Because the photoelectrons are strongly attenuated by passage through the sample material itself, the information obtained comes from the sample surface, with a sampling depth on the order of 5 nm. Chemical bonding in molecules will cause binding energy shifts, which can thus be used to extract information of a chemical nature (such as atomic oxidation state) from the sample surface.

In practice, the electron binding energy (E_b) is determined as follows. Because the energy of an X-ray with particular wavelength is known (e.g., Al K$_\alpha$ X-ray has a photon energy of $E_{ph} = 1,486.7$ eV) and the kinetic energy of the emitted electrons (E_{kin}) can be measured, E_b of each of the emitted electrons can be determined by using the following equation:

$$E_b = E_{ph} - (E_{kin} + \phi) \tag{3.5}$$

where ϕ is the work function dependent on both the spectrometer and the material. This equation is basically for conservation of energy. The work function ϕ is an adjustable instrumental correction factor that accounts for the few eV of kinetic energy given up by the photoelectron as it becomes absorbed by the instrument detector, which rarely needs to be adjusted in practice.

XPS has been used frequently to characterize materials including MHPs. For example, Figure 3.18 shows the XPS survey spectra $CH_3NH_3PbI_3$ perovskite films for fresh (taken after 25 hrs of the sample preparation) and aged samples (after 500 hrs and 1,000 hrs) [40]. The elements identified in the samples include indium, tin, titanium, lead, iodine, oxygen, nitrogen, and carbon. The presence of carbon, nitrogen, lead, and

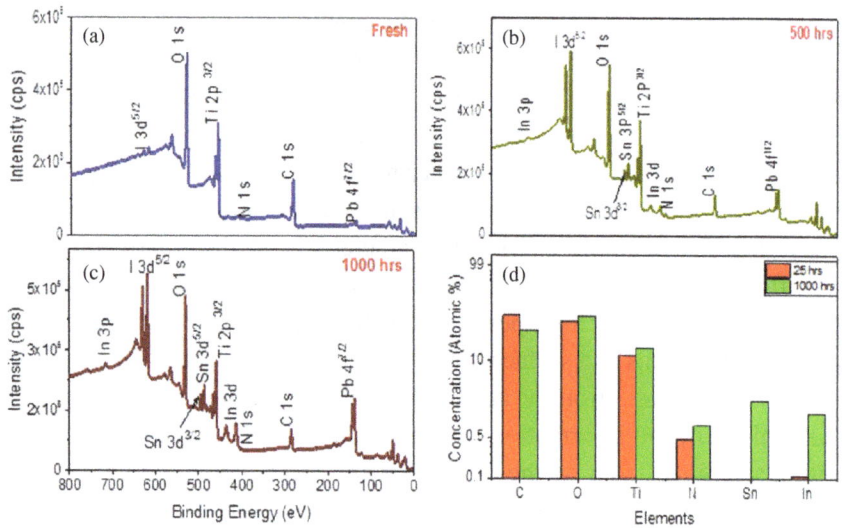

Figure 3.18. XPS survey spectra of the $CH_3NH_3PbI_3$ perovskite films documented for (a) fresh samples (taken after 25 hrs of sample preparation) and after the aging of (b) 500 hrs and (c) 1,000 hrs, in open air under room temperature. The graph (d) shows a comparison of the concentrations (in atomic %) of C, N, O, Ti, In, and Sn between fresh and aged (1,000 hrs) perovskite layers. Reproduced from Ref. [40]

iodine are expectedly based on the chemical composition of the $CH_3NH_3PbI_3$ perovskite film.

XPS has been also used to study how perovskites interact with metal oxide substrates in solar cells. For instance, Figure 3.19 shows high resolution XPS spectra and their chemical or compositional changes with time or aging in open air of $CH_3NH_3PbI_3$ perovskite films on indium tin oxide (ITO) substrate [40]. The features at the 413.2 and 435.2 eV correspond to indium metal and In_2O_3, respectively. The binding energies of 485.6, 486.2, and 487 eV of the Sn $3d^{5/2}$ core spectrum are attributed to tin, tin oxide, and halide, respectively. The moisture absorbed by the perovskite film was found to play a dominant role in accelerating the migration of indium and tin from the ITO layer, where the migration rate depends on the moisture level in the air and the exposure period. The core level spectrum of lead (Figure 3.19(c)) shows two distinct peaks at Pb $4f^{5/2}$ (~143 eV) and Pb $4f^{7/2}$ (138 eV), with the fresh sample showing the existence of PbO_2 and $PbCO_3$ and increase in lead oxides with aging.

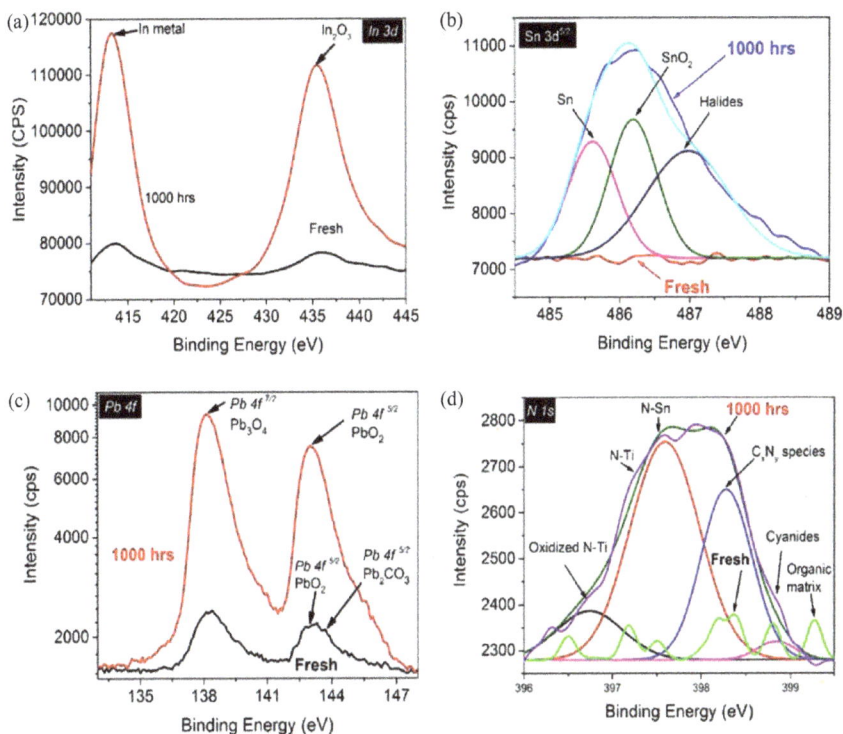

Figure 3.19. Core level XPS spectra (of (a) In 3d, (b) Sn $3d^{5/2}$, (c) Pb 4 f, and (d) N 1s) for fresh sample and 1,000 hrs aged samples. The In 3d spectra indicate significant increase of In metal content and formation of In_2O_3 at the surface of the perovskite layer after aging for 1,000 hrs. Sn $3d^{5/2}$ spectra show the existence of the Sn after 100 hrs. Also, the formation of the SnO_2 and halides is very clear in the aged sample. Pb 4 f spectra clearly indicates that the PbO_3 and PbO_2 have been significantly increased in the aged sample. N1s spectra exhibit the formation of metal nitrides in the aged sample. Reproduced from Ref. [40].

The electronic properties of solids are determined by or related to their electronic band structures that can be measured experimentally or calculated computationally. Electronic band structure is commonly studied by angular resolved photoelectron spectroscopy (ARPES or ARPS), which gives the joint density of states (DOS) from the valence bands (VBs) [41]. When photons with energy $h\nu$ are used to reject electrons from the surface with an emission angle of θ, which can be adjusted experimentally, the band dispersions $E(\boldsymbol{k})$ could be obtained simply from the measured values of kinetic energy (E_{kin}) based on equation as follows:

$$E_{kin} = h\nu - \phi - E_b \tag{3.6}$$

$$\hbar k_{\parallel} = \sqrt{2mE_{kin}} \times sin\theta \tag{3.7}$$

$$\hbar k_{\parallel} = \sqrt{2m(h\nu - \phi - E_b)} \times sin\theta \tag{3.8}$$

$$\hbar k_{\perp} = \sqrt{2m(E_{kin}cos^2\theta - V_0)} \times sin\theta \tag{3.9}$$

where $\hbar k_{\parallel}$ and $\hbar k_{\perp}$ are the components parallel and perpendicular to the surface of the electron crystal momentum, V_0 denotes the band depth from vacuum and is also called the *inner potential*, accounting for the electron work function φ, and the value of V_0 can be determined by examining only the electrons emitted perpendicular to the surface and measuring their kinetic energy as a function of incident photon energy. If we know ϕ or V_0 from literature or control experiments, for each angle θ chosen, one can experimentally measure E_{kin}, and then determine E_b for a given photon energy $h\nu$. Using the above equations, one can calculate $\hbar k_{\parallel}$ and $\hbar k_{\perp}$, and thus also $\hbar \boldsymbol{k} = \hbar k_{\parallel} + \hbar k_{\perp}$. The equations for energy and momentum can be solved together to determine the dispersion relation between the binding energy, E_b, and the wave vector, $\boldsymbol{k}$, of the electron.

For instance, Figure 3.20 shows a comparison between calculated and experimentally observed band structures for MAPbBr$_3$ based on ARPES in the X–M and X–R momentum directions of the bulk Brillouin zone (BZ) [42]. By overlaying the calculated bands along the X–M and X–R directions with the photoemission data, good overall agreement was found between theory and experiment. For example, in experiment, the bandwidths found for MAPbBr$_3$ amount to 1.3 and 0.7 eV along the X–R and X–M directions, respectively, which compare well with the calculated bandwidths (1.5 and 0.8 eV for these two directions).

With detailed knowledge of the single-crystal band structures of the two MHPs, one can understand the electronic structure of polycrystalline films and determine the VB onset. Using MAPbI$_3$ as an example, energy

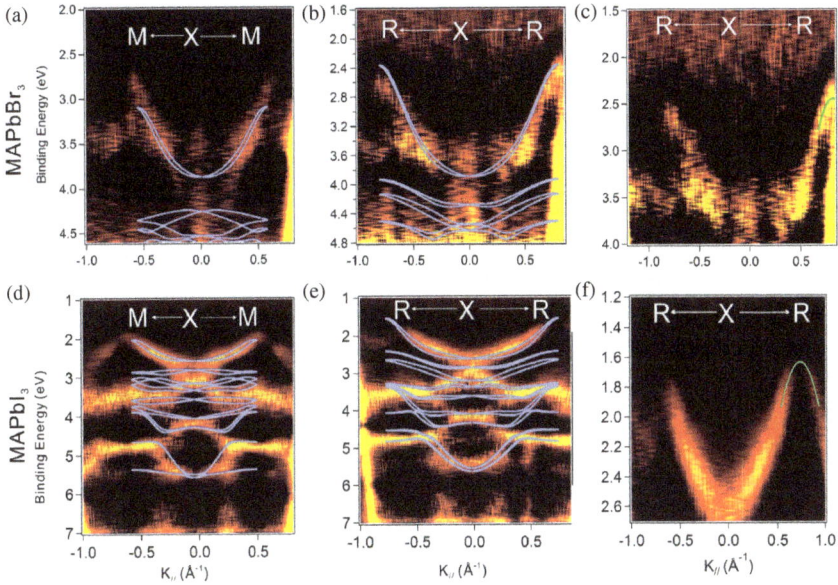

Figure 3.20. ARPES *k*-space 2D curvature band maps. Top row: MAPbBr$_3$ single crystal along the (a) X–M and (b) X–R high-symmetry directions. Bottom row: MAPbI$_3$ single crystal along the (d) X–M and (e) X–R high-symmetry directions. In all plots, correspond-ing DFT-calculated bands are shown with light blue lines (shifted in energy to match experimental band positions). A zoom into the topmost VB along the X–R direction is given in (c) for MAPbBr$_3$ and in (f) for MAPbI$_3$. Fitting the VB edge regions with para-bolic curves (green lines) yields a hole effective mass near R of ~0.25 ± 0.05 m_0 for MAPbBr$_3$ and ~0.50 ± 0.10 m_0 for MAPbI$_3$. Reproduced from Ref. [42].

distribution curves (EDCs), i.e., photoemission intensity as a function of electron binding energy (BE) for defined emission angles at different $k_\parallel$ values, have been used to determine the VB onset, as shown in Figure 3.21, plotted in both linear and logarithmic scales [42].

Theoretical studies play a critical role in understanding the complex band structures of MHPs. Figure 3.22(a) shows the energy band diagram derived from quasiparticle self-consistent GW Approximation (QSGW) calculations along the relevant *k*-branches defined on Figure 3.21(b) [43]. The energy difference between conduction and VB is plotted in Figure 3.15(c) for different band pairs. Extrema of the curves in recipro-cal space, where the conduction and VB energies are close to being

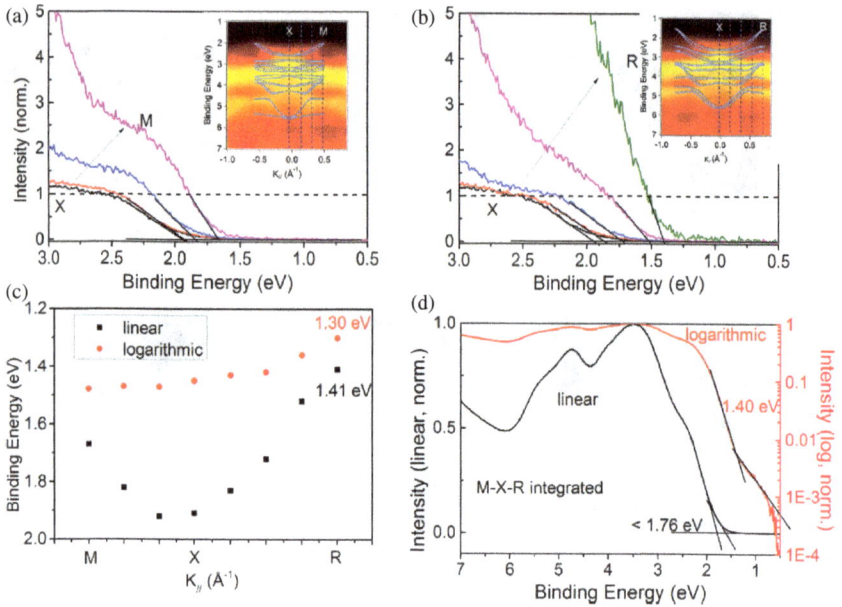

Figure 3.21. Energy distribution curves (EDC) and valence band maximum (VBM) determination for MAPbI$_3$. (a, b) EDCs at selected k values along the X–M and X–R directions, respectively. The corresponding k values for the EDCs are indicated by the vertical dashed lines in the ARPES k-space maps (insets in (a) and (b)). Spectra are normalized at the corresponding top VB peak positions in order to better illustrate the k dependence of the VB onsets. (c) VBM determination from the EDCs plotted in (a) and (b), using linear and logarithmic intensity scales for selected k values along M–X–R. (d) Simulated MAPbI$_3$ polycrystalline thin film UPS spectrum, obtained by summation of all EDCs from (a) and (b), shown on linear and logarithmic scales to highlight the variation in VBM determination. Reproduced from Ref. [42]

parallel, corresponding to critical points (CPs) of the joint DOS. The joint DOS is the number of electronic states per unit energy and volume in the conduction and VBs that are separated by a given photon energy, taking into account spin degeneracy and selection rules. Points where the conduction and VBs track each other (i.e., where the gradient $\nabla_k(E_c - E_v)$ is minimal) dominate the dielectric function. This is replotted in Figure 3.15(d) together with the relative contributions from different inter-band transitions.

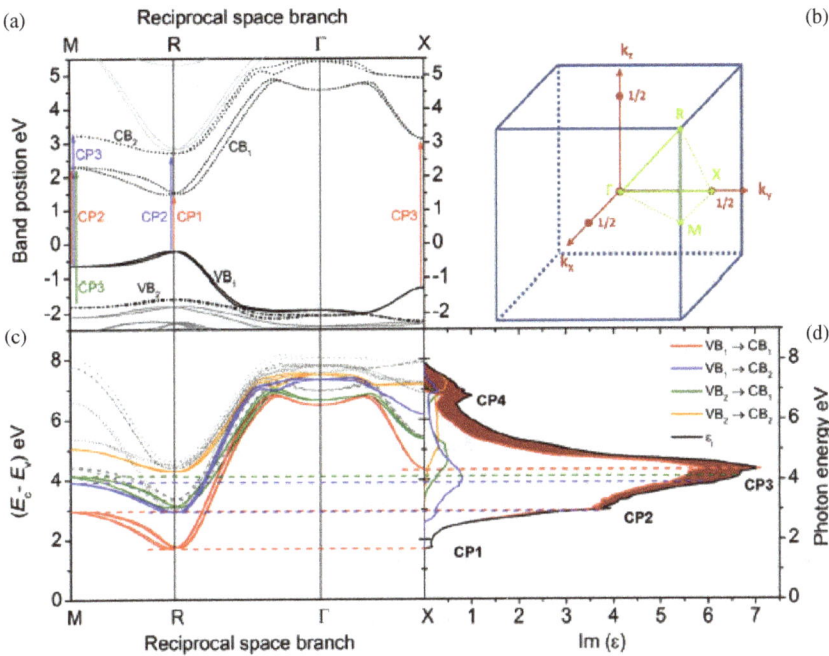

Figure 3.22. (a) Energy band diagram of MAPbI$_3$ along the relevant k-branches in the first Brillouin zone. The labels tag the critical point transitions with the appropriate inter-band transitions. (b) Schematic representation of the first Brillouin zone showing the position of the high symmetry points used in *a* and *c*. (c) Difference between conduction and valence band energy plotted for relevant band pairs. The color code is the same as given in the inset to *d*. (d) Imaginary part of the dielectric function ($Im(\varepsilon)$). The color code helps visualizing the contributions of the different transitions to $Im(\varepsilon)$. The area filled with dark red shows the small contribution to the imaginary part of the dielectric function from other transitions than VB$_1$ to CB$_1$. The lines between *c* and *d* underline the correspondence between the extrema of $E_c - E_v$ and the points of high absorption. (e) Simulated transient absorption spectrum for a charge carrier density of 10^{-5} per unit cell (4×10^{16} cm^{-3}). The dashed black line corresponds to the summed contribution of all transitions and is rescaled according to the procedure detailed in the Methods section. The main individual contributions are shown (solid colored lines). "PB$_1$," and "PB$_2$," label the two photo-bleaches. The inset shows schematic band diagrams and optical transitions for three different scenarios: (i) no photo-excited population, (ii) photo-excited population following a pulse with energy greater than 2.6 eV (477 nm), and (iii) after excitation with a pulse energy between 1.6 and 2.5 eV (780 and 500 nm) or following relaxation of photoexcited population from (ii) within the CBs and VBs. Reproduced from Ref. [43].

(e)

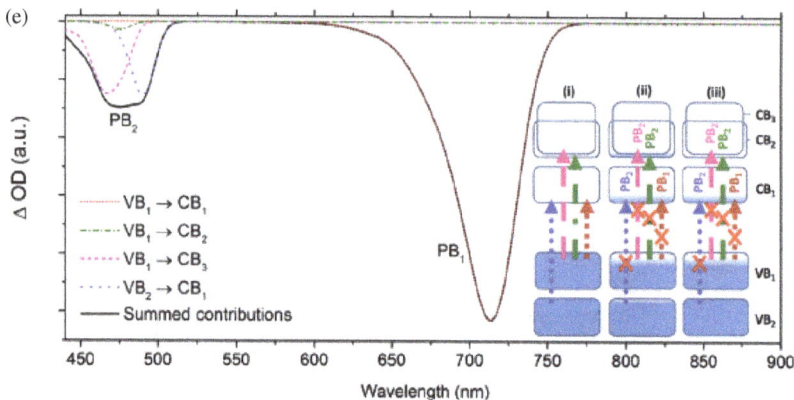

Figure 3.22.   (*Continued*)

3.2.2 *Charge Carrier Mobility and Conductivity*

In solid materials, charge carriers, e.g., electrons and holes, can move in response to an applied external electrical field. Electron mobility characterizes how quickly an electron can move through a metal or semiconductor, when subject to an electric field. Likewise, hole mobility is used to describe how quickly a hole moves when the material is under an electrical field. In general, charge carrier mobility refers to both electron and hole mobility.

Electron and hole mobilities are special cases of electrical mobility of charged particles in a fluid under an applied electric field. When an electric field E is applied across a piece of material, the electrons respond by moving with an average velocity called the *drift velocity, v_d*. The electron mobility μ is defined as

$$v_d = \mu \, E \tag{3.10}$$

Electron mobility is usually given in the unit of $cm^2/(V \cdot s)$, which is different from the SI unit of mobility, $m^2/(V \cdot s)$. They are related by 1 $m^2/(V \cdot s) = 10^4 \, cm^2/(V \cdot s)$.

Charge carrier mobility is directly related to electrical conductivity. If n is the number density (concentration) of electrons and μ_e is their mobility, in the electric field E, each of these electrons will move with the velocity vector $-\mu_e E$, for a total current density of $ne\mu_e E$ (where e is the elementary charge). Therefore, the electrical conductivity σ satisfies

$$\sigma = ne\mu_e \tag{3.11}$$

This formula is valid when the conductivity is due entirely to electrons. In a p-type semiconductor, the conductivity is due to holes (h) instead, but the formula is essentially the same: If p is the concentration of holes and μ_h is the hole mobility, then the conductivity is

$$\sigma = pe\mu_p \tag{3.12}$$

If a semiconductor has both electrons and holes, the total conductivity is

$$\sigma = ne\mu_e + pe\mu_p \tag{3.13}$$

For example, typical electron mobility for Si at room temperature (300 K) is 1,400 cm^2/(V·s) and the hole mobility is around 450 cm^2/(V·s). Therefore, the conductivity depends on both density and mobility of the charge carriers. When other things are equal, higher mobility leads to higher conductivity and better device performance.

Semiconductor mobility depends on the impurity concentrations (e.g., donor and acceptor concentrations), defect concentration, temperature, and electron and hole concentrations. It also depends on the electric field, particularly at high fields when velocity saturation occurs.

Charge carrier mobility is usually determined using the Hall effect or inferred from transistor behavior measured using field effect transistors (FET). There are other techniques that have been used to measure carrier mobility, including time of flight (TOF), terahertz spectroscopy, and PL quenching (PLQ) [44].

In Hall effect measurement, the Hall voltage (V_H) and current (I) are measured for a sample with a thickness of l (usually thin films and, e.g., in the x and y plane) subject to an external magic field (B) perpendicular to the sample (z direction). The electron mobility (μ_n) is given by

$$\mu_n = -\frac{\sigma_n V_{Hn} l}{I B} \tag{3.14}$$

and, similarly, for the hole mobility

$$\mu_p = \frac{\sigma_n V_{Hp} l}{I B} \tag{3.15}$$

Once the values of V_{Hn} or V_{Hp}, l, I, and B are measured, the electron or hole mobility can be calculated.

Another common approach to measuring carrier mobility is using FET. In this case, the measurement can be done either in a saturation mode or a linear region. In the saturation mode, the carrier mobility is given by

$$\mu = m_{sat}^2 \frac{2L}{W} \frac{1}{C_i} \tag{3.16}$$

where L and W are the length and width of the channel, C_i is the gate insulator capacitance per unit area, and m_{sat} is determined from the slope of plotting the square root of the saturated current (I_D), reached by increasing the drain-source voltage V_{DS}, against the gate voltage V_{GS}.

In the linear region (Ohmic mode), the carrier mobility is determined by

$$\mu = m_{sat}^2 \frac{L}{W} \frac{1}{V_{DS}} \frac{1}{C_i} \tag{3.17}$$

where V_{DS} is small, I_D is linearly proportional to V_{GS} with slope m_{lin}.

The mobility of charge carriers in MHPs has been studied extensively and found to depend strongly on the characteristics of the samples, as for other solid materials. It is influenced by a number of both intrinsic and extrinsic factors, such as impurity scattering, phonon scattering, and surface scattering. In particular, the morphology and crystallinity have a strong effect on mobility. The value also varies somewhat with measurement techniques that have different errors or approximations made in analysis.

For example, for MAPbI$_3$, the mobility for films fabricated from spin coating or crystal growth from solution varies from 0.7 to 1.4 cm^2/(V s) for the electron, from 0.4 to 0.9 cm^2/(V s) for the hole, and from 31–75 cm^2/(V s) for the sum of the electron and hole [45]. For single crystals, the mobility is higher, in the range of 2.5 to 164 cm^2/(V s) for the hole, and 115 to 600 cm^2/(V s) for the sum of the electron and hole. Similar large variations have been found for other MHPs [45]. The large variation has been attributed to a number of possible factors, including difference in measurement techniques, variation in the sample quality and thickness, or approximations or assumptions made in analysis.

The crystallinity and overall structure and shape are also critical. For example, the diffusion lengths in CH$_3$NH$_3$PbI$_3$ single crystals grown by a solution-growth method can exceed 175 μm under 1 sun (100 mW cm^{-2}) illumination and exceed 3 mm under weak light for both electrons and holes [2]. These long diffusion lengths result from a combination of greater carrier mobility (25 cm^2/(V s for the electron and 105 cm^2/(V s for the hole based on Hall measurement), longer lifetime (~100 μs under one sun), and much smaller trap densities (~10^{10} cm^{-3}) in the single crystals than in polycrystalline thin films. In another study, the intrinsic Hall mobility and photocarrier recombination coefficient are directly measured in CH$_3$NH$_3$PbI$_3$ and CH$_3$NH$_3$PbBr$_3$ in steady-state transport studies [46]. The results show that electron–hole recombination and carrier trapping rates in these perovskites are very low. Figure 3.23 shows the setups and results of charge carrier mobility characterization [46].

3.3 Optical Properties

One of the major attractive properties of MHPs is their strong optical absorption in the visible region of the spectrum, which is important for many applications including PV and LEDs. The strong absorption is due to the direct bandgap (dipole allowed) nature of the first excitonic absorption and large oscillator strength determined by the dipole moment operator and the wave functions of the CB and VB involved. Strong absorption is not only important for PV but also for applications such as light detection.

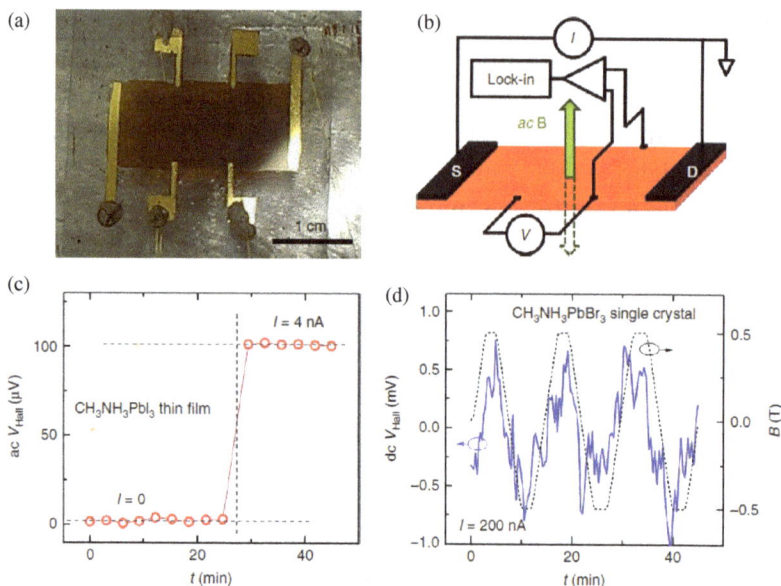

Figure 3.23.   Hall effect measurements in hybrid lead–halide perovskites. (a) A photo of typical solution-grown $CH_3NH_3PbI_3$ thin film on glass with Au contacts in a 4-probe/Hall bar geometry. (b) A diagram of the a.c. Hall effect measurement setup. (c) A representative Faraday-induction-corrected a.c. Hall measurement in a solution-grown $CH_3NH_3PbI_3$ thin film shown in (a). When a d.c. excitation current of 4 nA is driven through the film uniformly illuminated with a blue light ($\lambda = 465$ nm) and subjected to an a.c. B field of r.m.s. magnitude 0.23 T at 0.5 Hz, an a.c. Hall voltage of 100 μV is detected above the zero-bias background by using a lock-in technique shown in (b). The vertical and horizontal dashed lines show the moment d.c. excitation current is turned on (at $t = 28$ min) and the two levels of the Hall voltage, at I = 0 and 4 nA, respectively. Note that even though Hall mobility of this sample is only $\mu_{Hall} = 1.5$ cm^2 V^{-1} s^{-1}, the signal-to-noise ratio is excellent (the standard deviation in V_{Hall} among consecutive six measurements is only about 0.5%). (d) A representative d.c. Hall measurement in $CH_3NH_3PbBr_3$ single crystals. V_{Hall} is measured with an electrometer at a constant excitation current I = 200 nA, while B field is slowly swept between –0.5 T and 0.5 T. Despite the much higher mobility of this sample ($\mu_{Hall} = 11\pm3$ cm^2 V^{-1} s^{-1}), the signal-to-noise ratio is much worse, with the large error imposed by the noise. Reproduced from Ref. [46].

In principle, the electronic absorption spectrum in the UV-visible region is determined by the electronic band structure of the material. In the semiclassical approach to spectroscopy, Fermi's Golden rule for dipole allowed transitions shows the electronic transition responsible for optical absorption is determined by

$$\langle \Psi_f | \mu \cdot E | \Psi_i \rangle \tag{3.18}$$

where Ψ_i and Ψ_f are the wavefunctions of the initial and final states, μ is the electrical dipole moment operator (a vector), and E is the applied external electrical field (also a vector), e.g., light. The braket notation implies complete integration over the entire space in all dimensions involved by the wavefunctions and dipole moment operator. If we consider E as space independent on the scale of the chromophore, it can be taken out of the integration or braket. The evaluation of Eq. (3.18) determines if the transition is allowed (if the integration does not result in zero) and how strong the transition is, i.e., the spectral intensity (proportional to the square of the value calculated by Eq. (3.18)).

For electronic absorption of a semiconductor, the initial state is the VB and the final state is the conduction band (CB). For PL, the initial state is the CB while the final state is the VB. Based on the Tauc law, for direct allowed transitions, the absorption coefficient is related to the band gap E_g by

$$\alpha(h\nu) = A^* (h\nu - E_g)^{1/2} \tag{3.19}$$

where A^* is a constant determined by the index of refraction and electron and hole effective masses [47,48]. For transitions that are only allowed at $k \neq 0$ but forbidden at $k = 0$, the absorption coefficient has the following expression:

$$\alpha(h\nu) = A'(h\nu - E_g)^{3/2} \tag{3.20}$$

where, similar to A^* in Eq. (3.19), A' is related to the index of refraction and effective masses of the electron and hole. The absorption spectrum for direct bandgap transitions often feature a clear excitonic band (e.g., CdS and MAPbI$_3$) while that for indirect bandgap transitions lacks a sharp band and features absorption intensity monotonically increasing toward shorter wavelength (e.g., Si).

MHPs are unique in that they have strong intrinsic absorption in the UV and visible regions, which, in conjunction with long charge carrier diffusion length or high mobility, is desired for PV applications. The strong absorption is a result of their electronic band structures. For example, Figure 3.24 shows a comparison between experimentally measured

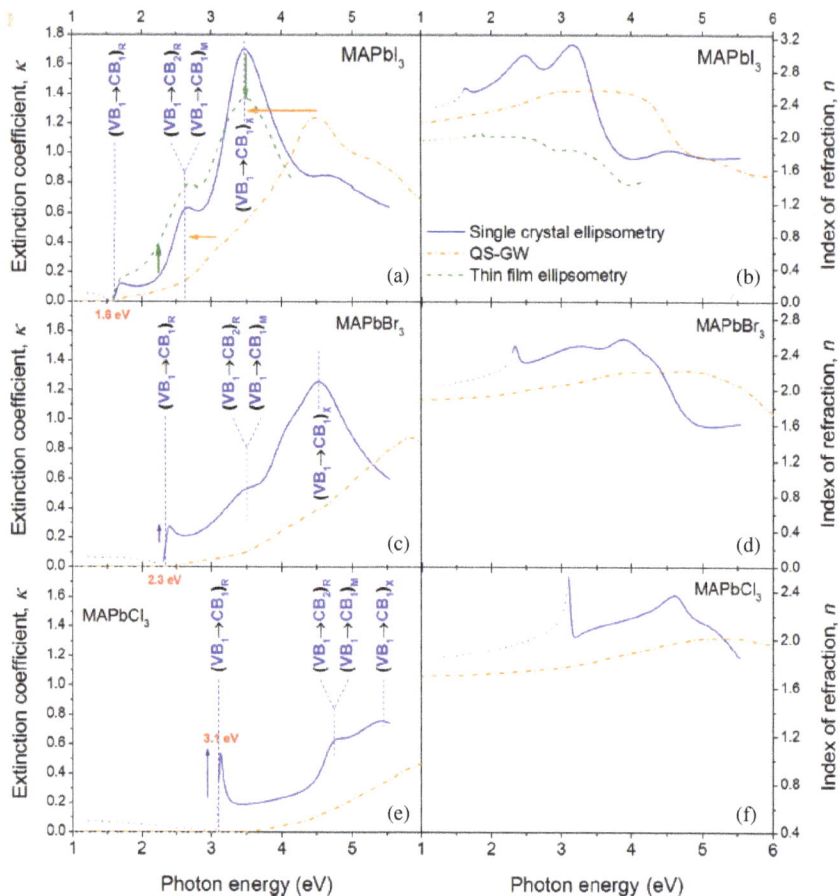

Figure 3.24. Optical constants (extinction coefficient (a, c, e) and index of refraction (b, d, f)) of $CH_3NH_3PbI_3$ (a and b), $CH_3NH_3PbBr_3$ (c and d), and $CH_3NH_3PbCl_3$ (e and f), derived from single crystal ellipsometry (solid blue lines), thin film $CH_3NH_3PbI_3$ (dashed green lines), and theory (chained orange lines, see Methods section). The three main optical features of $CH_3NH_3PbI_3$ are assigned to the appropriate inter-band transition (VB and CB stand for "valence band" and "conduction band", respectively. The subscripts "1" and "2" correspond to the highest (lowest) VB (CB) or second highest (lowest) VB (CB)). The subscripts outside the brackets designate the symmetry point where the transition occurs in the first Brillouin zone, as defined in Figure 3.21. Reproduced from Ref. [43].

and computational calculated optical properties (extinction coefficients and index of refraction) of MAPbX$_3$ [43]. Substituting iodide by lighter halides increases the bandgap of the compound from ~1.55 (I) to 2.24 (Br) to 2.97 (Cl) eV. Three main absorption peaks (corresponding to critical points or CPs) can be observed in MAPbI$_3$ at 1.6, 2.5, and 3.1 eV. In MAPbBr$_3$, three peaks can also be seen at ~2.3, 3.5, and 4.5 eV. MAPbCl$_3$ shows only two peaks within the measured range, at ~3.1 and 4.6 eV. The absorption peak near the band gap increases in amplitude for lighter halides, as indicated by the blue arrows. This behavior is likely to be due to excitonic absorption. Detailed assignment of the optical transitions, made in conjunction with Figure 3.16, has been provided in Ref. [43].

MHPs have generally shown high PLQY with exitonic or bandedge emission for both single crystals, polycrystalline films, and nanostructures. The high PLQY indicates a low density of bandgap trap states. In some cases, trap state emissions red-shifted relative to excitonic emission are also observed. There is also strong interest in generating broad PL for white light generation, which relies on some sort of trap state emission and thus usually has lower PLQY. For example, as shown in Figure 3.25,

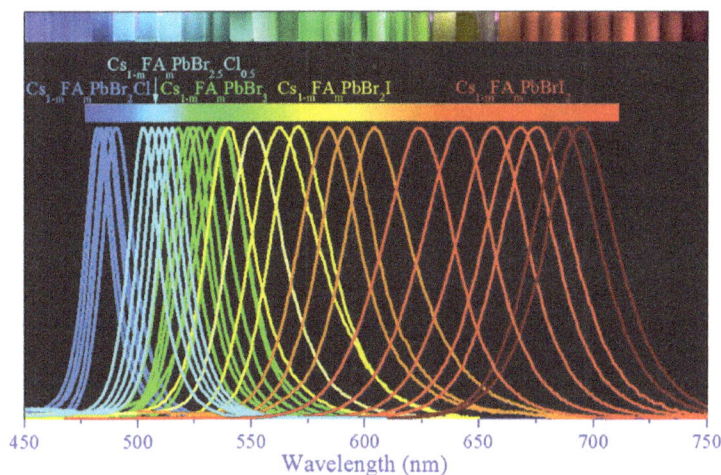

Figure 3.25. Full-spectral fine-tuning visible emissions from hybrid Cs$_{1-m}$FA$_m$PbX$_3$ (X = Cl, Br, I, 0 < m < 1) QDs with the help of concurrent cation and anion substitution. Insets (top) show the corresponding luminescent photographs upon irradiation with a UV (375 nm) lamp. Reproduced from Ref. [49].

highly tunable PL has been demonstrated in $Cs_{1-m}FA_mPbX_3$ (X = Cl, Br, I, $0 < m < 1$) QDs by changing the ratio between Cs and FA and altering the halide ions [49]. Such broad tunability is unusual and desirable for many applications such as optical sensing and imaging.

Similarly, MHPs have strong electroluminescence (EL) that is important for LED applications. For instance, room-temperature EL is demonstrated from a light-emitting diode (LED) based on two-dimensional lead halide perovskites, $(C_6H_5NH_3)_2PbI_4$, with some represented data shown in Figure 3.26 [50]. The device has a multilayer structure of ITO/

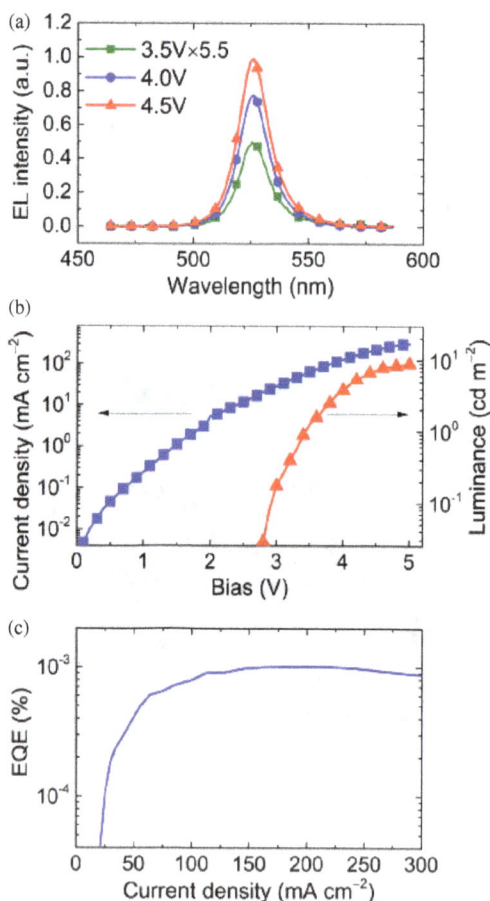

Figure. 3.26. (a) EL spectrum of the $(C_6H_5CH_2NH_3)_2PbI_4$ perovskite LED (PeLED) under 3.5 V ($\times$ 5.5), 4 V, and 4.5 V. (b) Current density (J) and luminance versus voltage (V) of the PeLED. (c) EQE versus J of the PeLED. Reproduced from Ref. [50].

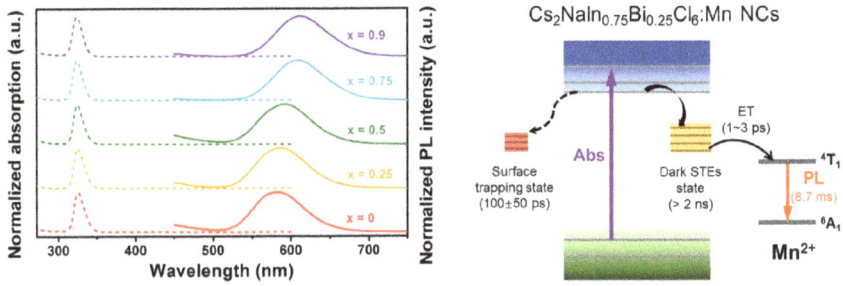

Figure 3.27. (Left) Normalized steady-state absorption (dashed line) and PL (continuous lines) spectra of Mn-doped $Cs_2NaIn_xBi_{1-x}Cl_6$ (x = 0, 0.25, 0.5, 0.75, and 0.9) NCs. (Right) Proposed exciton dynamics model to explain energy transfer and PL emission. Reproduced from Ref. [51].

Poly(3,4-ethylenedioxythiophene): Poly(styrenesulfonate)/$C_6H_5NH_3)_2PbI_4$/Bathophenanthroline/Al. The EL emission peaks at 526 nm with a narrow FWHM of 15 nm, which origins from perovskite exciton emission. The LED device exhibits a maximum luminance of 9 cd m^2 at a bias of 5 V [50].

Chemical doping is a common approach to alter the electronic, optical, or magnetic properties of semiconductors including MHPs. For light emission applications of MHPs with ABX_3 structures, in particular, both A and B sites can be used to introduce dopants, while the X component can be varied from I to Br or Cl. For example, Mn^{2+} has been used to dope $Cs_2NaIn_xBi_{1-x}Cl_6$ (x = 0, 0.25, 0.5, 0.75, and 0.9) NCs, with their steady-state absorption (dashed line) and PL (continuous lines) spectra shown in Figure 3.27, along with a proposed exciton dynamics model to explain energy transfer and associated PL emission [51]. Many other dopants have been studied including Ag^+, $Cu^{2+/1+}$, Ni^{2+}, Bi^{3+}, and different rare earth metal ions [19].

3.4 Nonlinear Optical Properties

MHPs also exhibit some unique and interesting nonlinear optical (NLO) properties that are promising for various photonics and imaging applications. Common NLO processes include second harmonic generation (SHG), sum-frequency generation (SFG), and higher-order processes such as third- and fourth-order harmonic generation. The NLO properties are strongly dependent on the characteristics of the materials involved.

Using SHG as an example, which is the most commonly studied and used, when light with angular frequency ω, whose electrical field is given by $E(\omega)$, is focused into a sample, light with the frequency of 2ω may be generated and detected and its electrical field $E(2\omega)$ is given as follows:

$$E(2\omega) \propto P^{(2)}(2\omega) = \chi^{(2)} E(\omega) E(\omega) \tag{3.21}$$

where $P^{(2)}(2\omega)$ is the induced second harmonic dipole per unit volume, and $\chi^{(2)}$ is the nonlinear susceptibility tensor, which is characteristic of the materials or molecules at surfaces or interfaces, such as symmetry and orientation. Crystals with centro-symmetry do not allow SHG. At the surfaces or interface of such crystals, SHG is allowed due to the broken symmetry. Therefore, SHG is often used as a surface sensitive technique for studying properties of molecules at surfaces/interfaces. For crystals with no centro-symmetry, SHG is allowed inside the crystals.

For example, 2D halide perovskites with relatively bulky organic A sites allow for the introduction of chiral amines, which could induce non-centro-symmetry required for intrinsic and bulk second-order NLO effects. Chiral β-methylphenethylamine (MPEA) has been used as the organic precursor to direct the non-centrosymmetric assembly of the 2D inorganic layers of the halide perovskites, generating perovskite nanowires with a chiral and non-centro-symmetric P1 space group [52]. The material is designated as $(MPEA)_{1.5}PbBr_{3.5}(DMSO)_{0.5}$, since DMSO solvent molecules coordinate axially with Pb^{2+} in the partially edge-shared octahedrons (Figure 3.28(a)). The displacement of halide Br with DMSO in the $[PbX_6]$ octahedrons and the resulted structure distortion cause the long-range polar nature of the chiral perovskite nanowires (Figure 3.28(b–e)). NLO measurements on the single nanowires demonstrated an effective second-order NLO coefficient (d_{eff}) of ≈ 0.68 pm V^{-1} (comparable favorably with that of potassium dihydrogen phosphate, KDP), with a polarization ratio as high as $(96.4 \pm 0.1)\%$ when pumped with linearly polarized fs laser pulse at 850 nm (Figure 3.28(g)). This MHP exhibited a good thermo-, air-, and optical stability, with a laser damage threshold of ≈ 0.52 mJ cm^{-2} (8.6×104 W cm^{-2}) (Figure 3.28(f)). These nanowires also exhibit an NLO circular dichroism (CD) when pumped with circularly

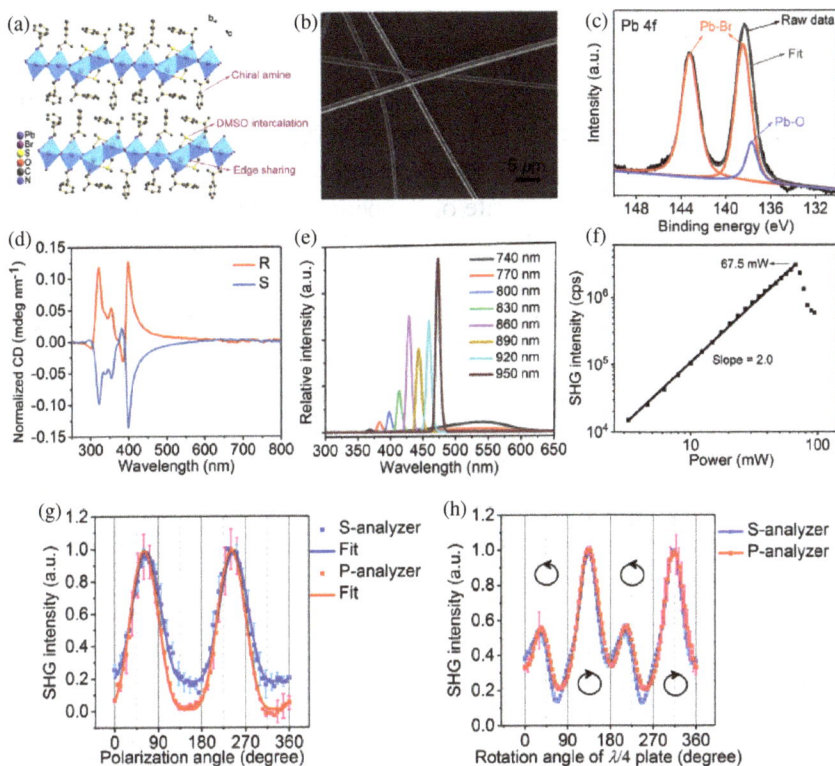

Figure 3.28. Chiral organic–inorganic halide perovskite for second-order NLO. (a) Chemical structure of chiral (R-MPEA)$_{1.5}$PbBr$_{3.5}$(DMSO)$_{0.5}$. (b) SEM image of the (R-MPEA)$_{1.5}$PbBr$_{3.5}$(DMSO)$_{0.5}$ nanowires. (c) Decomposed high-resolution XPS spectrum of Pb 4f for the (R-MPEA)$_{1.5}$PbBr$_{3.5}$(DMSO)$_{0.5}$ nanowires. (d) Normalized CD spectra of the films spin-coated from DMF solutions of the R- and S-perovskite crystals. (e) NLO spectra of a (R-MPEA)$_{1.5}$PbBr$_{3.5}$(DMSO)$_{0.5}$ nanowire pumped at various wavelengths. (f) Logarithmic plot of SHG intensity as function of the incident power. (g) Linear polarization dependence of the SHG intensity from a single chiral perovskite (R-MPEA)$_{1.5}$PbBr$_{3.5}$(DMSO)$_{0.5}$ nanowire. The dots are the experimental data, while the solid lines are the cos$^4\theta$ fits. (h) Circular polarization dependence of SHG intensity from the nanowire as function of the rotation angle of the λ/4 plate. The pump was left-handed circularly polarized when the rotation angle was 45° and 225° and was right-handed circularly polarized when the rotation angle was 135° and 315°, as indicated by the arrows. Reproduced from Ref. [53].

polarized light, with an SHG-CD of $(74.0 \pm 0.1)\%$ when the analyzer is perpendicular to the nanowire axis (Figure 3.28(h)).

Another interesting NLO property is amplified spontaneous emission (ASE) and related lasing. ASE and lasing have been observed in MHPs. For instance, continuous wave (CW) lasing has been demonstrated in a mixed cation system at the A site of the perovskite crystal structure, $Cs_{0.1}$ $(MA_{0.17}FA_{0.83})_{0.9}Pb_{0.84}(I_{0.84}Br_{0.16})_{2.68}$ at any temperature up to 120 K [54]. Figure 3.29 shows the spontaneous emission and ASE spectra as a function or pump (excitation) fluence (Figure 3.29(a)–3.29(c)) and ASE threshold as a function of temperature (Figure 3.29(d)) [54]. The temperature-dependent ASE threshold was measured under 0.8 ns pulsed excitation, and the threshold level for CW ASE is 387 W cm^{-2} at 80 K. The results indicate that easily fabricated single-phase perovskite thin films can sustain CW stimulated emission at low temperature. There is potential for operating at higher temperatures upon optimization of the perovskite materials.

3.5 Magnetic Properties

MHPs generally do not possess strong intrinsic magnetic properties, but can exhibit magnetic properties due to defects or doping. For example, defect mediated room temperature ferromagnetism (FM) in MA lead halide perovskite has been observed experimentally and supported by *ab initio* calculations [55]. Theoretical analysis suggests that the FM mainly arises from iodide vacancies in the orthorhombic and cubic crystal phases but not in the tetragonal phase. The ferromagnetic hysteresis loop is stable in the 100–380 K range, which substantiates the notion that the magnetism is defect related.

Doping is a common approach to alter the electronic, optical as well as magnetic properties of semiconductors including MHPs. For example, Mn^{2+} doping or substitution for Pb^{2+} in $MAPbI_3$ dramatically modifies the magnetic properties of the system as observed by electron spin resonance (ESR) measurements performed in an exceptionally broad 9–315 GHz frequency range (Figure 3.23). While the pristine material is non-magnetic, with only ppm level of paramagnetic impurities detected, Mn substitution gives a strong ESR signal, indicating FM, which is also

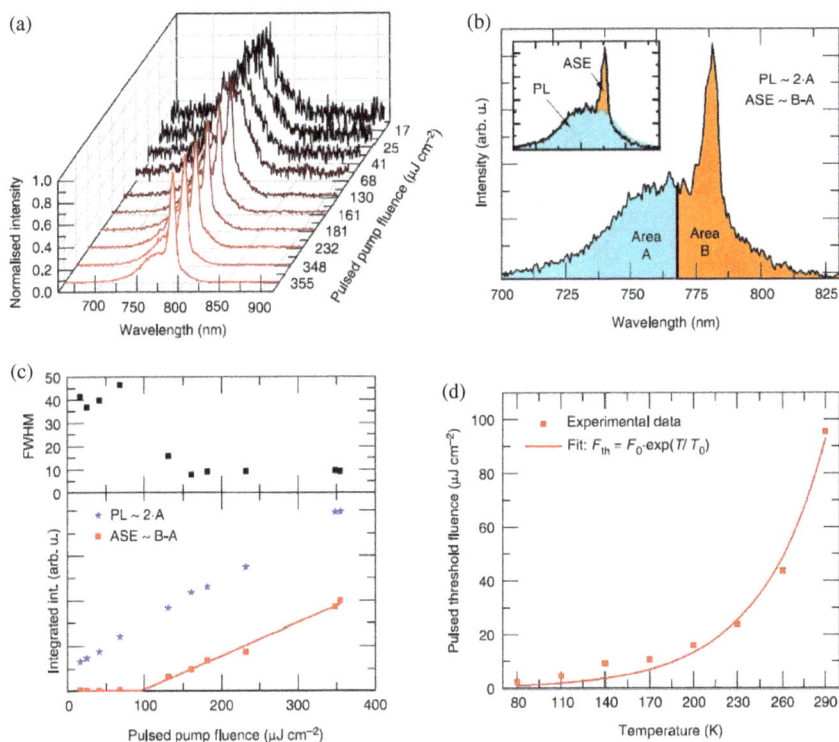

Figure 3.29. Amplified spontaneous emission (ASE) properties under pulsed excitation (~0.8 ns). (a) Example of data for room temperature showing the evolution of the emission spectra from which the ASE threshold is established. Different colors for the spectra are chosen for better distinguishability. (b) Regions of the spectrum used to determine spontaneous emission (PL) and ASE contributions. The inset graphically shows the assigned PL contribution (blue, area A and area A mirrored at the center wavelength) and ASE contribution (orange, area B–area A), and the real spectrum (black line). (c) PL, ASE, and full width at half maximum (FWHM) as a function of the pulse fluence. (d) ASE threshold as a function of temperature. The threshold decreases by more than an order of magnitude from room temperature to 80 K. The red line is a fit to the empirical exponential threshold law typically observed in inorganic semiconductors. The characteristic temperature T_0 is found to be 46 K. Reproduced from Ref. [54].

supported by SQUID measurements (Figure 3.30(a)) [56]. Importantly, a ferromagnetic order develops below $T_C = 25$ K on cooling in dark, and the ferromagnetic order causes a rapid shift of the resonant field, B_0, and the broadening of the linewidth, ΔB, below T_C, which are sensitive measures of the magnetic interactions and the internal magnetic fields. To help understand the origin of the FM, first-principles calculations of the atomic configurations and magnetic order have been conducted to determine the total DOS and projected density of states (PDOS) for the in-plane model of $CH_3NH_3(Mn:Pb)I_3$ in its neutral FM configuration (Figure 3.30(b)), along with some structural information related to the Pb–I and Mn–I bonds (Figure 3.30(c) and 3.30(d)). It was also demonstrated that high efficiency photocurrent generation by low power visible light illumination results in a melting (from ordered to disordered) of the ferromagnetic state and a small local field can set the direction of the magnetic moment. This offers a simple and efficient way of optical spin control and has implications in applications such as low power, light controlled magnetic devices.

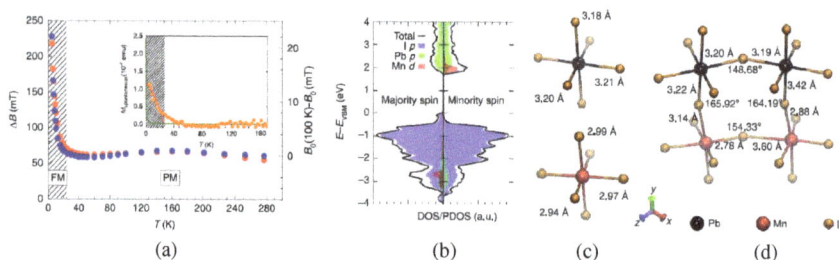

Figure 3.30.   Magnetic properties of $CH_3NH_3(Mn:Pb)I_3$ in dark. (a) ESR linewidth (red dots) and resonant field (blue dots, offset by a reference value B_0) as a function of temperature recorded at 9.4 GHz. Their temperature-independent behavior is a characteristic of the paramagnetic phase (PM). The upturn below 25 K corresponds to the on-set of the ferromagnetic (FM) phase. Inset: SQUID magnetometry of MAMn:PbI$_3$. The temperature dependence of the spontaneous magnetization measured in 1.2 μT magnetic field shows a clear increase below T_C. The red line represents the $M_0(1 - (T/T_C)^{3/2})$ temperature dependence given by Bloch's Law. (b) First-principles calculations of the atomic configurations and magnetic order show total density of states (DOS) and projected density of states (PDOS) calculated for the in-plane model of $CH_3NH_3(Mn:Pb)I_3$ in its neutral FM configuration. (c) Calculated Pb–I and Mn–I distances for a single Mn dopant. (d) Calculated bond angles and bond distances for the I mediated super-exchange paths in the FM ground state of the in-plane model of $CH_3NH_3(Mn:Pb)I_3$. Reproduced from Ref. [56].

3.6 Unique Properties of Nanostructured Metal Halide Perovskites

3.6.1 *Ensembled Systems*

While nanostructures of MHPs share many similar properties to their bulk counterparts, they do possess some unique features compared to bulk. The first is the much larger surface-to-volume (S/V) ratio. The second is the quantum confinement (QC) effect, especially when the physical dimension becomes comparable to or smaller than the Bohr exciton radius.

For example, for a spherical nanoparticle with radius R, the S/V is proportional to $1/R$. There is an increase in S/V ratio of about one million when R is decreased from 1 cm to 10 nm. The extremely large surface area has many implications to the properties and functionalities of the nanostructures, including perovskites as well as other semiconductors and metals.

Again, assuming a spherical semiconductor nanoparticle with radius R, the effective bandgap, $E_{g,eff}(R)$, is given by

$$E_{g,eff}(R) = E_g(\infty) + \frac{\hbar^2 \pi^2}{2R^2} \left(\frac{1}{m^*_e} + \frac{1}{m^*_h} \right) - \frac{1.8e^2}{\varepsilon R} \tag{3.22}$$

where $E_g(\infty)$ is the bulk bandgap, m^*_e and m^*_h are the effective masses of the electron and hole, and ε is the bulk optical dielectric constant or relative permittivity. The second term on the right-hand side shows that the effective bandgap is inversely proportional to R^2 and increases with decrease in size. In contrast, the third term due to Coulombic interaction leads to a decrease with bandgap with decreasing R. However, since the second term becomes dominant at small R, the effective bandgap will increase with decreasing R, especially when R is small, the so-called quantum confined regime. This has been predicted and observed in many semiconductor QDs or NCs. Figure 3.31 shows an illustration of the size dependence of the energy levels, e.g., from single atoms through clusters and QDs to bulk.

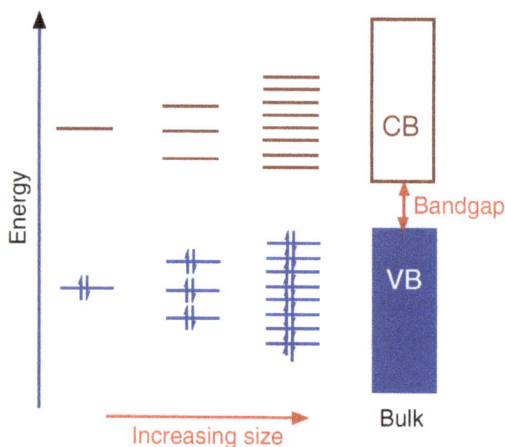

Figure 3.31. Illustration of size-dependent electronic energy levels for a semiconductor, from single atoms or "molecules" through clusters and QDs to bulk.

QC effect has also been demonstrated in MHP QDs or NCs. For example, Figure 3.32 shows the absorption and PL spectra of different sized $CsPbBr_3$ NCs and a comparison between the experimentally determined and theoretically calculated size dependence of the bandgap [57]. With decreasing size of the NCs, both the absorption and PL excitonic peaks blue shift due to increased bandgaps. Therefore, the color of the NCs can be tuned at will by changing the NC size. This is desired for various optical or photoelectronic applications.

The optical properties or color of the MHPs can also be changed by altering the chemical composition, e.g., changing A, B, or X in the ABX_3 structure. Figure 3.33 shows an example of changing the halide or varying the ratio of mixed halides to dramatically change the color or electronic absorption and PL spectra in a wide spectral range [57]. In changing the halide ions, the color is redder for heavier halide. This color change is what one would expect based on the "heavy atom" effect that suggests, if other factors are equal or similar, that heavier atoms with a higher density of electronic states and smaller energy level spacing lead to redder absorption and emission of light (longer wavelength and lower energy). Similar findings have been made for $MAPbX_3$ [58]. Such color tunability is highly desired for applications such as imaging and LEDs [58].

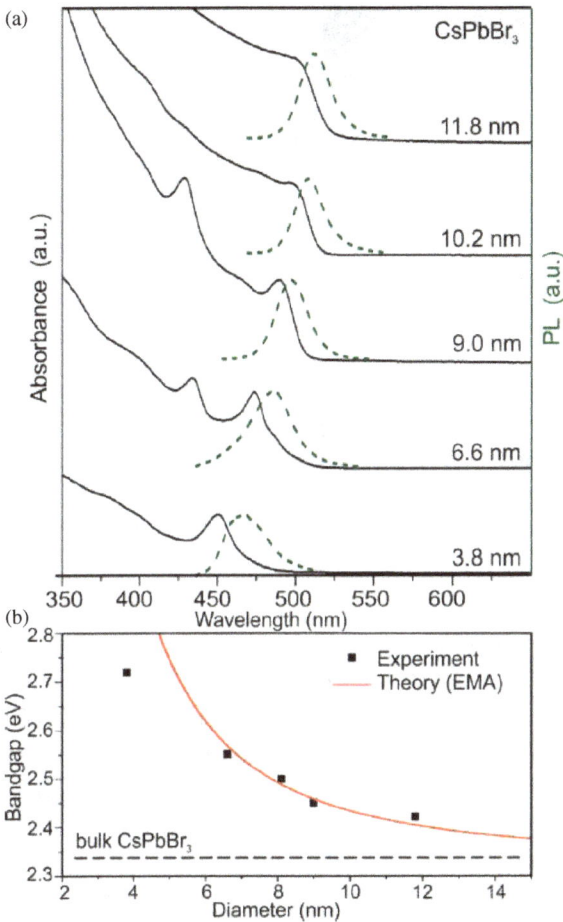

Figure 3.32. (a) Quantum size confinement effect in the absorption and emission spectra of 5–12 nm CsPbBr$_3$ NCs. (b) Experimental versus theoretical (effective mass approximation, EMA) size dependence of the bandgap energy. Reproduced from Ref. [57].

Another example of changing composition is to change A or B in the ABX$_3$ MHP structures. Since the electronic absorption and PL are dominated by band edge states that are mainly related to B and X, changing A does not dramatically change the bandgap or optical properties but changing B does have a major effect, similar to changing X. For example, replacing Pb with Sn has some effect on the electronic absorption and PL spectra as shown in Figure 3.34. One noticeable change is the much larger

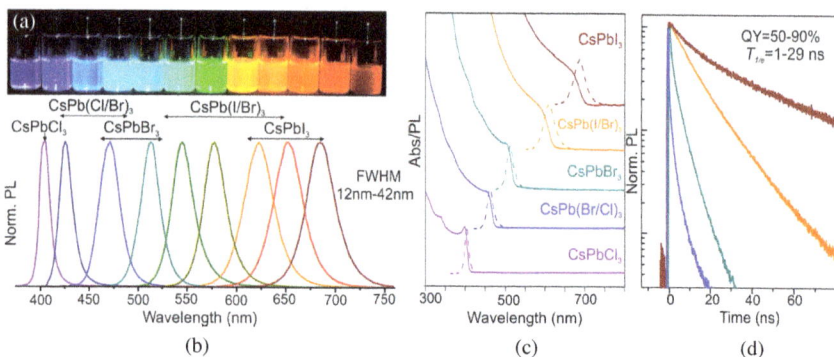

Figure 3.33.   Colloidal perovskite $CsPbX_3$ NCs (X = Cl, Br, I) exhibit size- and composition-tunable bandgap energies covering the entire visible spectral region with narrow and bright emission: (a) colloidal solutions in toluene under UV lamp ($\lambda = 365$ nm), (b) representative PL spectra ($\lambda_{exc} = 400$ nm for all but 350 nm for $CsPbCl_3$ samples), (c) typical optical absorption and PL spectra; (d) time-resolved PL decays for all samples shown in (c) except $CsPbCl_3$. Reproduced from Ref. [57].

Stokes shift between the PL peak and first excitonic absorption peak, which reflects the difference in band structures between $CsPbX_3$ and $CsSnX_3$. In this example, the authors pointed out that the PL spectrum of $CsSn(Cl_{0.5}Br_{0.5})_3$ NCs was identical to the pure chloride-containing NCs but did not give a reason as to why. This observation is somewhat surprising since the UV visible electronic absorption spectra between the two samples are different, with the first excitonic absorption peak around 420 nm for $CsSnCl_3$ and 510 nm for $CsSn(Cl_{0.5}Br_{0.5})_3$ [59], since usually PL peak positions shift with excitonic absorption peak positions.

3.6.2 *Doped Perovskite Nanostructures*

Another way to alter and control the optical properties of MHPs is to dope with one or more other elements to replace a small percentage of the A or B component of the ABX_3 structure. For example, Figure 3.35 shows Mn-doped $CsPbCl_3$ PQDs with an increased PLQY, measured by the characteristic Mn^{2+} PL band centered near 580 nm, as the amount of Mn^{2+} dopant, replacing Pb^{2+}, increases while the bandedge PL from the host $CsPbCl_3$ remains about the same in PLQY [60]. The doping does not

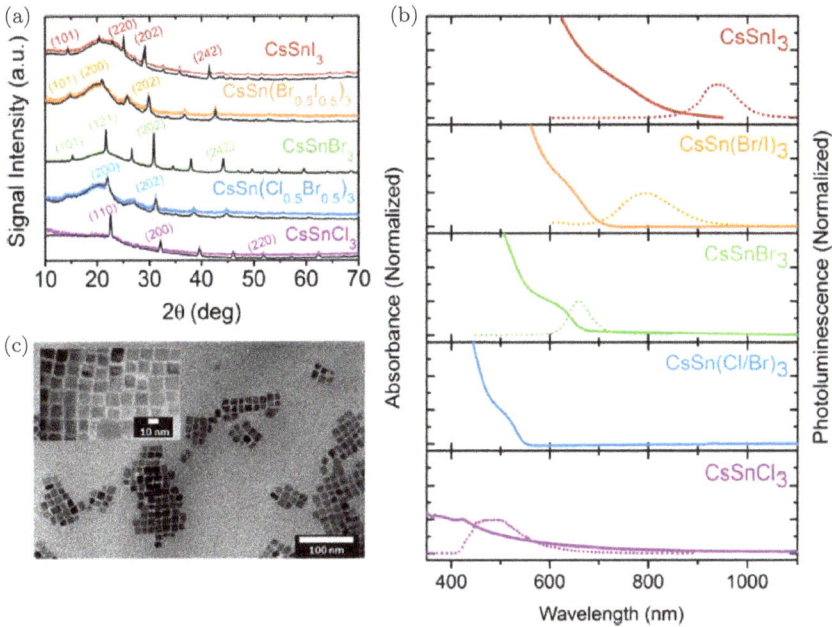

Figure 3.34. (a) PXRD spectra of $CsSnX_3$ (X = Cl, $Cl_{0.5}Br_{0.5}$, Br, $Br_{0.5}I_{0.5}$, I) perovskite nanocrystals. (b) Absorbance and steady-state PL of nanocrystals containing pure and mixed halides. The PL spectrum of $CsSn(Cl_{0.5}Br_{0.5})_3$ particles was identical to the pure chloride-containing nanocrystals and is shown in the SI S4. (c) TEM image of $CsSnI_3$ nanocrystals. Reproduced from Ref. [59].

affect the excitonic absorption as expected since the absorption spectrum is dominated by the host and usually not affected by doping.

Other dopants have been studied as well for doping MHPs, including Ag^+, $Cu^{2+/1+}$, Bi^{3+}, and Ni^{2+}. While usually doping mainly affects photoluminescence and to a lesser degree absorption, in some situations, doping can result in significant changes in the bandgap of the host and thereby the optical absorption or color. For example, as shown in Figure 3.26, doping of $MAPbBr_3$ crystals with various Bi results in substantial redshift absorption as a result of bandgap narrowing, especially lowering of the conduction band edge [19].

Besides the B site, the A site can also be replaced or doped with other monovalent cations, e.g., alkali metals such as Na, K, and Cs to replace MA in $MAPbX_3$ [61]. For example, K^+ and Rb^+ have been introduced into

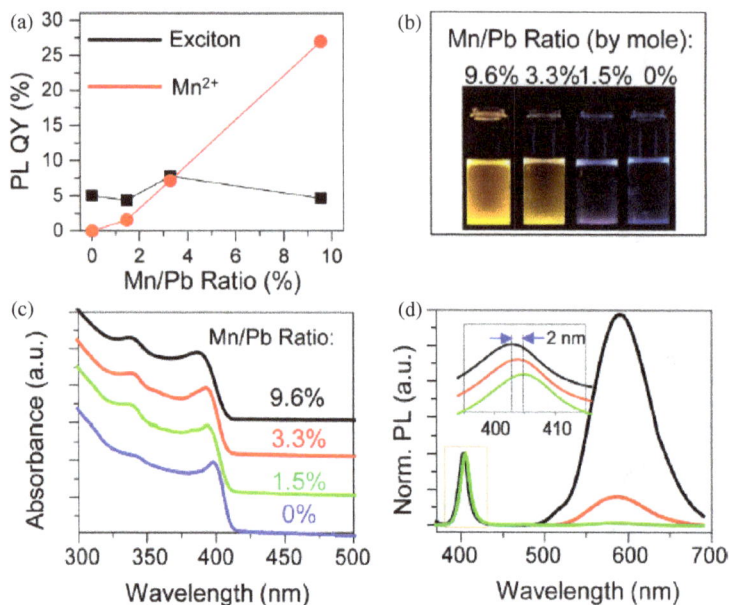

Figure 3.35. The dependence of Mn:CsPbCl$_3$ NC optical properties on dopant concentration. (a) While the absolute PL QY of the band-edge emission is largely independent of dopant concentration, the PL QY of the Mn^{2+} d–d transition depends on it directly. (b) Photograph of hexane solutions of Mn:CsPbCl$_3$ NCs of varying Mn-content illuminated by a UV lamp (365 nm). Solutions were diluted to exhibit the same optical density at 365 nm. The UV emission of the host NC is essentially invisible, while the orange Mn-based emission grows in intensity with dopant concentration. (c) Normalized absorption and (d) PL spectra of Mn:CsPbCl$_3$ NCs of varying dopant concentrations. Increased Mn-content is associated with a gradual blueshift of both the band-edge peak in absorption and the intrinsic NC PL peak (expanded in the inset of panel d); this can be attributed to the effects of alloying on the NC band structure. The Mn-emission peak grows in intensity with increasing Mn-content but does not shift. Reproduced from Ref. [60].

FA$_{1-x}$MA$_x$PbX$_3$ 2D perovskites, replacing FA or MA, to increase crystallinity and improve film morphology with less defects, which lead to enhanced PCE of perovskite solar cells [62].

3.6.3 *Properties of Single Perovskite Nanostructures*

Study of a single nanostructure, similar to that of a single molecule, allows one to remove some of the complications related to heterogeneity due to

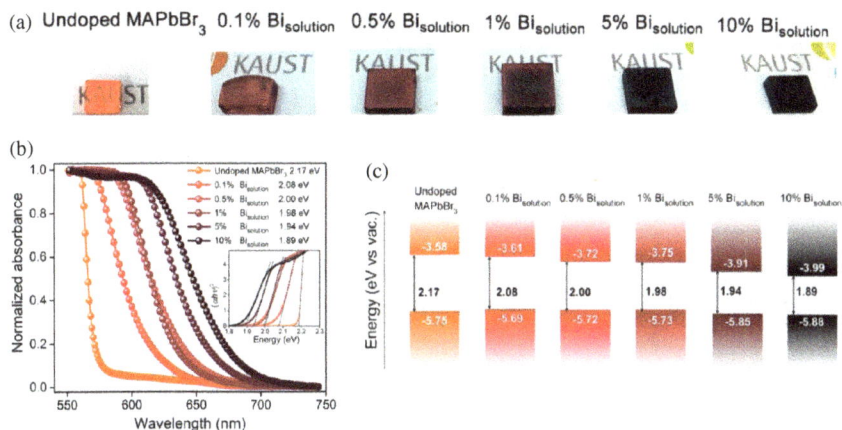

Figure 3.36. (a) Photographs showing MAPbBr$_3$ crystals with various Bi incorporation levels. (b) Steady-state absorption spectra of MAPbBr$_3$ crystals with various Bi%. Inset: corresponding Tauc plots. (c) Bandgap alignment of MAPbBr$_3$ crystals with various Bi%. Reproduced from Ref. [19].

variations in the size, shape, and surface of nanostructures. Single particle spectroscopy has been applied to study single perovskite nanostructures to gain insights, which are not possible in ensembled averaged samples. For example, single-particle spectroscopy has been used to study and compare EL and PL at the individual NC level for CsPbBr$_3$ perovskite, as shown in Figure 3.37 [63]. The NCs form aggregates in a conducting matrix used as an emission layer in an EL device. Only a small fraction of the NCs within the aggregate emits light as a result of efficient charge migration, accumulation, and selective recombination on larger NCs, leading to pronounced blinking and reduced efficiency (illustrated in Figure 3.37(k) and (l)). Under similar conditions of excitation rates in both EL and PL, the intrinsic QY of EL is on average lower than that of PL (about 36%).

In another study, single particle spectroscopy has been used, in conjunction with atomic force microscopy (AFM), to study the photodegradation process of CsPbBr$_3$ PNCs at the single-particle level by monitoring the size change [64]. Decreasing particle size and blueshift of PL of a single CsPbBr$_3$ PNC are observed upon continuous photoirradiation, as shown in Figure 3.38. This helps in gaining insights into the issue of instability of perovskites.

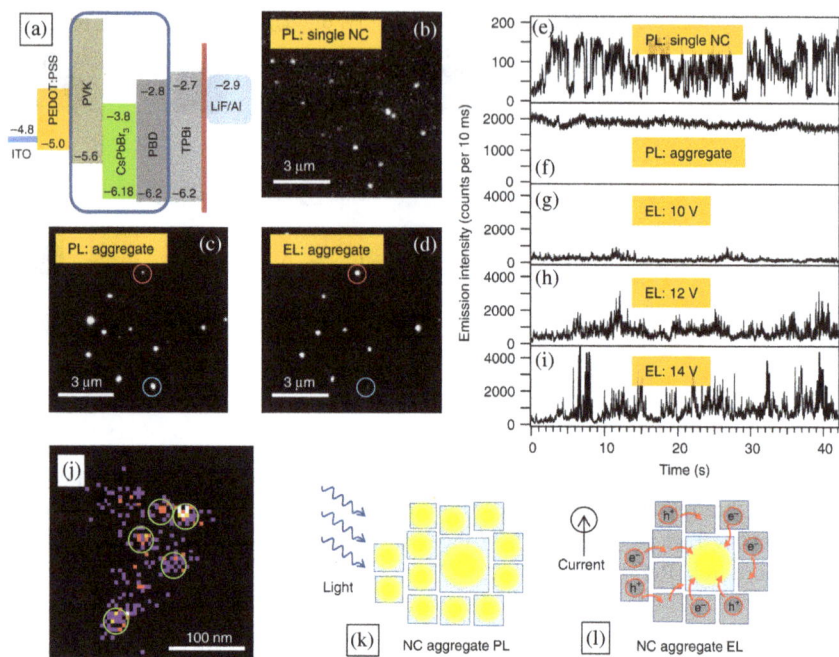

Figure 3.37.   Imaging and blinking of NC aggregates in PL and EL. (a) Scheme of the EL device structure, together with the relevant energy levels. (b) Microscopic PL image of single NCs of CsPbBr$_3$ in PVK matrix. (c) Microscopic PL image of aggregates of CsPbBr$_3$NCs in EL device. (d) Microscopic EL image (at a bias of 14 V) of the same NC aggregates as in (c). In (c) and (d), the blue circles denote an example of an aggregate which shows stronger PL, red circles, an aggregate which shows stronger EL. (e) Intensity time trace of PL of single CsPbBr$_3$NCs in PVK matrix. (f) Intensity time trace of PL of aggregates of CsPbBr$_3$NCs in EL device; intensity time trace of EL of aggregates of CsPbBr$_3$NCs at (g) 10; (h) 12; and (i) 14 V. (j) Super-resolution EL image obtained by analysis of a series of microscopic EL images of aggregates of CsPbBr$_3$NCs at 14 V over the time of 60 s. (k) Schematic illustration of PL from an aggregate showing independent emission from individual NCs. (l) Schematic illustration of EL from an aggregate showing charge migration (red arrows) and emission from the largest NC which works as a trap. Reproduced from Ref. [63].

3.7 Exciton and Charge Carrier Dynamics

Complementary to static or equilibrium studies, dynamics studies can help to gain a deeper understanding of some fundamental properties of materials. Exciton and charge carrier dynamics in MHPs have been

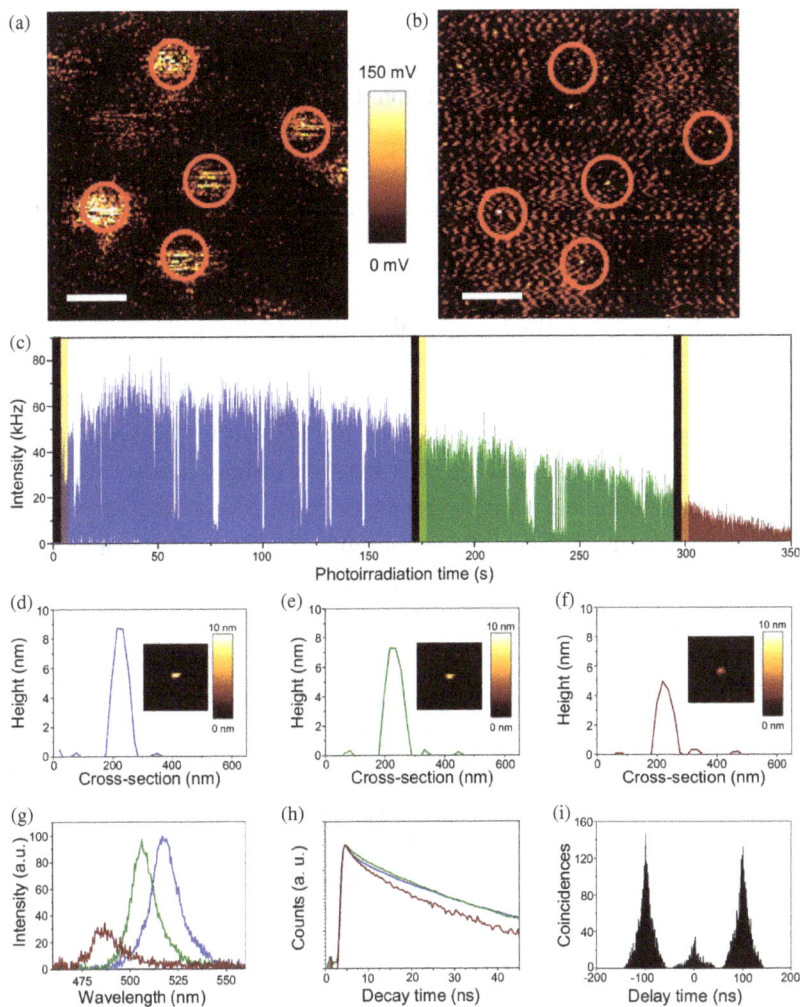

Figure 3.38. PL (a) and AFM topography (b) images of single PNCs obtained by simultaneous measurements. The scale bars in the images represent 1 μm. (c) PL time trace of a single PNC upon continuous photoirradiation. AFM images and PL spectra were measured at appropriate times, as indicated by black and yellow markers, respectively. Cross-sections and AFM images (inset) of the single PNC observed before photoirradiation (d), after 170 s photoirradiation (e), and after another 120 s photoirradiation (f). (g) PL spectra measured before continuous photoirradiation (blue line), after 170 s photoirradiation (green line), and after another 120 s photoirradiation (red line). These spectra were measured with a 5 s exposure time. (h) PL decay curves of the single PNC detected from the blue (blue line), green (green line), and red regions (red line) in (c). (i) Photon correlation histogram of the single PNC detected from the blue region in (c). Reproduced from Ref. [64].

studied in the form of bulk, polycrystalline films, and QD or NCs. The goal is often to learn more about the material properties beyond what is possible with static or frequency domain studies such as UV-visible, PL, TEM, and XRD.

Two common time-resolved techniques for dynamics studies are transient absorption (TA) and time-resolved photoluminescence (TRPL) [65]. TA is based on a pump-probe set up in which a short laser pulse (ps or fs) is used to excite (pump) the sample while a second short laser pulse, delayed with respect to the first one in time and lower in energy as not to perturb the system too much, is use to probe or interrogate the sample following its excitation. For example, the first pulse excites an electron from the VB of a semiconductor into the CB while the second pulse probes the excited electron in the CB by monitoring its absorption from the CB to a higher energy band. The experiment is repeated very quickly for different time delays between the pump and probe pulses using a translation stage, and the dynamics of the electron in the CB is reflected by the time profile measured. Since three states or bands are involved and not all are well known, interpretation of TA data is usually complicated [65].

TRPL involves exciting the sample with a short laser pulse like in TA but, instead of probing the dynamics with a second laser pulse based on absorption, the PL from the excited sample is detected by a detector that can resolve the PL in time, often on the ps and ns time scales. Shorter TRPL time resolution, e.g., fs, can be achieved using techniques such as PL up-conversion in which an optical gating method is used by mixing a second laser pulse with the PL and detecting the upconverted pulse light signal. This is a very involved technique and much less commonly used compared to TA or normal TRPL. Since only the ground and excited states (e.g., VB and CB) are involved, the interpretation of TRPL data is simpler than that of TA data.

Both TA and TRPL have been used to study exciton and charge carrier dynamics in MHPs. The exciton and charge carrier dynamics are in principle determined by the electronic band structures of the materials. In practice, there are bandgap states due to internal or surface defects that can also affect the dynamics, through processes such as trapping and trap state mediated recombination. Figure 3.39 shows several possible general dynamic processes for semiconductors such as MHPs [44], including

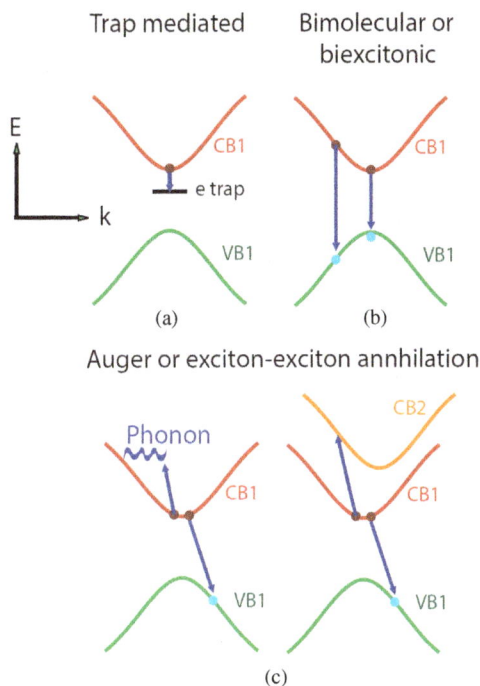

Figure 3.39. Schematic diagram indicating recombination mechanisms active in organic–inorganic metal halide perovskites. (a) Trap-assisted recombination is a monomolecular process involving the capture of either an electron (as shown) or a hole in a specific trap state (e.g., defect). (b) Bimolecular recombination may occur between electrons and holes, from either the relaxed state (CBM→VBM) or states higher in the band. (c) Auger recombination is a higher-order process involving at least three particles. The energy of an electron (or hole) is here transferred to another electron (or hole) to allow non-radiative recombination with a hole (or electron). As indicated, all processes have to satisfy energy and momentum conservation. Abbreviations: CB, conduction band; CBM, conduction band minimum; VB, valence band; VBM, valence band maximum. Reproduced from Ref. [44].

trap-assisted recombination as a monomolecular process involving the capture of either an electron or a hole in a specific trap state, "bimolecular" recombination between electrons and holes, from either the relaxed state (CBM→VBM) or states higher in the band, and Auger recombination as a higher-order process involving at least three particles with the energy of an electron (or hole) transferred to another electron (or hole) to allow non-radiative recombination with a hole (or electron). All the processes must satisfy energy and momentum conservation. What is not shown in this figure is another process called exciton–exciton annihilation, which involves at least two excitons or electron–hole pairs, or four particles, in which two excitons interact and one transfers energy to another, resulting in one being annihilated and another excited to generate an excited or hot exciton that ultimately decays by other mechanisms (radiative or non-radiative). The annihilation itself is non-radiative. Thus, exciton–exciton annihilation results in lower overall PLQY [66].

Figure 3.40 shows a theoretical analysis of the optical absorption spectra of bulk $CH_3NH_3MX_3$ based on group theory, electronic band diagram, and Fermi surface [67]. The electronic band structures are useful for understanding dynamics of exciton and charge carriers. Bandgap states are not considered in this analysis; thus, this may not represent the real situation. Experimentally, there will surely be defects for bulk sample, polycrystalline films, or especially nanostructures, which often complicates the interpretation of experimental results, both static, such as spectroscopic, and dynamic properties.

One important and interesting factor that can affect dynamics is spin-orbit coupling (SOC). As illustrated in Figure 3.41, SOC can be particularly important for experimental measurements that are spin sensitive. Indeed, recent studies have shown that spin affects the charge carrier dynamics related to the optical and magnetic properties of MHP, indicating potential applications in LEDs and spin-optotronics. In some cases, the organic component is replaced with an inorganic metal such as Cs. For example, bright triplet excitons are observed in $CsPbX_3$ (X = I, Br, Cl), which is attributed to the favorable effect of SOC in conjunction with broken inversion symmetry [68]. Similarly, high spin-polarization and ultra-large photoinduced magnetization of $CH_3NH_3PbI_3$ (MAPbI₃) have been achieved when circularly polarized light is used to excite the sample [69]. Figure 3.41(a) shows the energy bands of MAPbI₃ at the R-point (point group symmetry representation) with their respective levels from vacuum (experimental). The dashed box indicates the bands of interest with SOC. As shown in Figure 3.41(c), a spin relaxation lifetime of 7 ps for the electron (τ_e) and 1 ps for the hole (τ_h) was observed, making the materials potentially useful for ultrafast spin switch applications. Figure 3.41(b) gives a detailed explanation of the spin relaxation time or

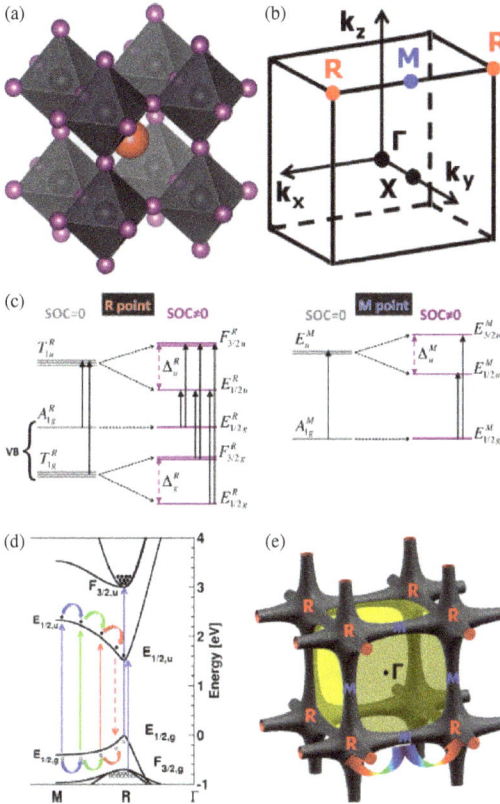

Figure 3.40. Analysis of optical absorption spectra of $CH_3NH_3MX_3$ based on group theory, electronic band diagram, and Fermi surface. (a) Real space 3D view of the Pm3m cubic crystal structure of metal–halide hybrid perovskites of general formulas $CH_3NH_3MX_3$. The $CH_3NH_3^+$ cation is located at the center of the cube with an averaged position sketched by a red ball. (b) Reciprocal space 3D view showing the first Brillouin Zone (BZ) of the Pm3m space group. Points of high symmetry in the cubic BZ are indicated by conventional letters: Γ denotes the origin of the BZ; X is the center of a square face at the BZ boundary, M is a center of a cube edge; and Rs are vertices of the cube. (c) Schematic energy level diagrams drawn at the high symmetry R and M points of the Pm3m BZ. Irreducible representations of the simple (SOC = 0, where SOC denotes spin orbit coupling) and double (SOC $\neq$ 0) groups are indicated. The level splitting associated with SOC are labeled Δ_g^R, Δ_u^R, and Δ_g^M. The additional spin degeneracy is not indicated for the double group representations. The upward arrows show the symmetry allowed optical transitions. (d) Electronic band diagram of the high temperature Pm3m cubic phase of C $CH_3NH_3MI_3$ taking SOC into account; an upward energy shift of 1.4 eV has been applied to match the experimental bandgap value at R. Carrier localization and transport after optical excitation are sketched. (e) Fermi surface ($E = -0.5$ eV) in the first BZ. R and M are connected along the edges, highlighting the saddle point nature of M. Reproduced from Ref. [67].

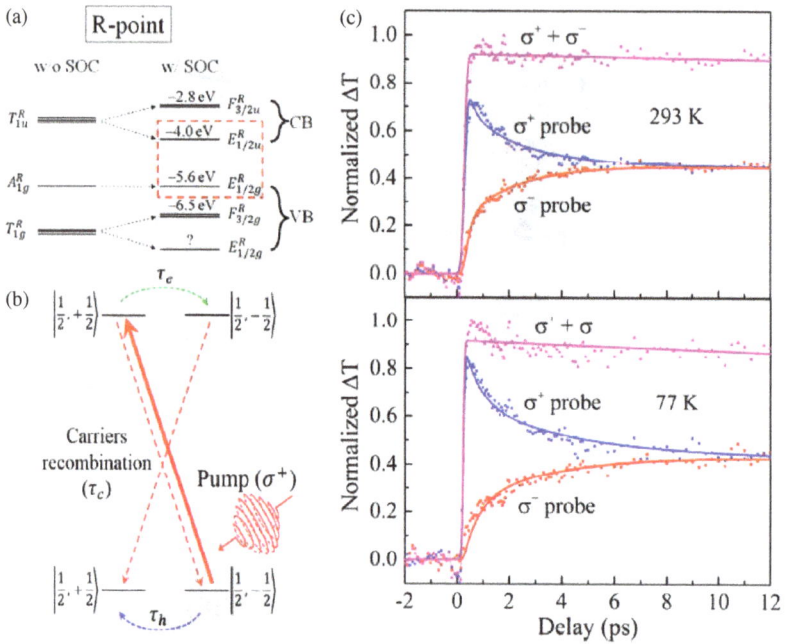

Figure 3.41. (a) Energy bands of $CH_3NH_3PbI_3$ at R-point (point group symmetry representation)[5] with their respective levels from vacuum (experimental).[2,3] Dashed box indicates the bands of interest. (b) Model of near band-edge photoexcitation by $\sigma+$ photon and J-states dynamics of $CH_3NH_3PbI_3$. The state notation is written as $|J,m_J\rangle$ where $J = 1/2$ is the electron's total angular momentum quantum number and $m_J = \pm 1/2$ is its projection in the z-axis. Absorption of σ^+ pump will raise the angular momentum by $+ \hbar$ ($\Delta m_J = +1$). (c) Normalized circular pump–probe decay transients with 19 $\mu J/cm^2$ σ^+ pump and σ^+ probe (blue), σ^- probe (red), and their total (magenta) at 293 K (top) and 77 K (bottom) for $CH_3NH_3PbI_3$ thin films. The experimental data are globally fitted for σ^+ probe, σ^- probe, and their sum. Reproduced from Ref. [69].

J relaxation time (intraband interstates transfer time or "J-flip" corresponding to spin flip). Due to strong spin-orbit coupling, or even Rashba and Dresselhaus couplings [70], MHPs are considered ideal for controlling spin states for spin-optotronics applications [71,72].

Dynamics of exciton and charge carriers have also been studied for nanostructures of MHPs such as PQDs and PNCs. For example, Figure 3.42 shows the transient bleach (TB) recovery profiles of PNCs passivated with octylammonium bromide (OABr) and octadecylammonium bromide

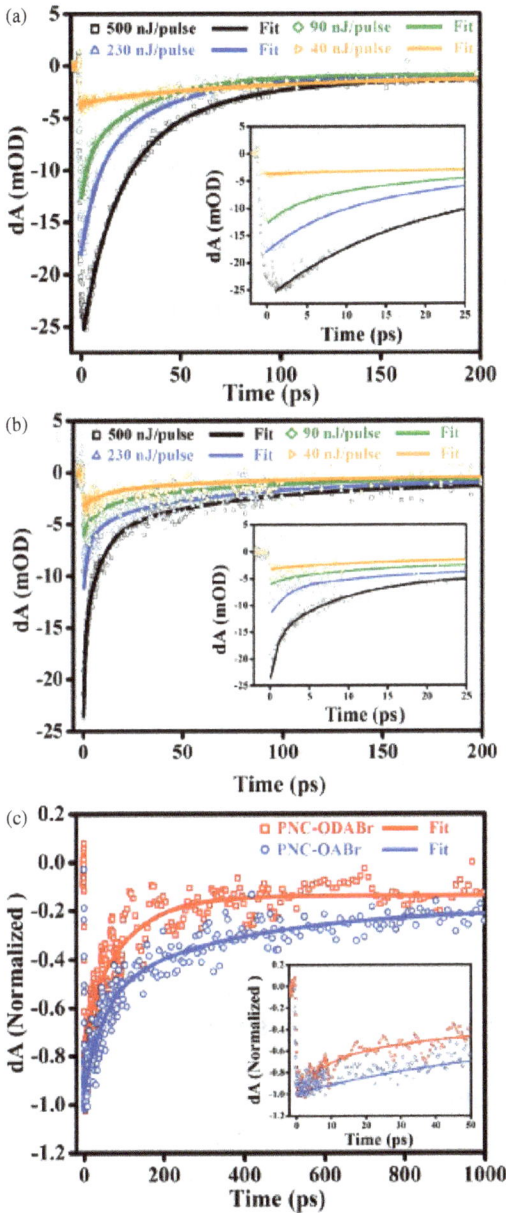

Figure 3.42. TB recovery profiles of (a) PNC–OABr and (b) PNC–ODABr with a pump power of 40, 90, 230, and 500 nJ/pulse over 0–200 ps time delay. (Inset) TB profiles over 0–25 ps time delay. (c) Normalized TB recovery profiles of PNC–OABr and PNC–ODABr with 40 nJ/pulse over 0–1,000 ps time delay. (Inset) TB profiles over 0–50 ps time delay. Reproduced from Ref. [73].

(ODABr), labelled as PNC–OABr and PNC–ODABr, under four different pump powers (40, 90, 230, and 500 nJ/pulse). The pump power dependence study is important for determining possible nonlinear processes that can complicate the interpretation of dynamics results, as explained further later. The TB recovery traces were fit with a triple-exponential function, and the amplitude of the fast component increased with increasing pump power nonlinearly (near quadratic) for both PNC samples, which is attributed to exciton–exciton annihilation [73]. The longer time (linear) dynamics are dependent on the passivating ligands and indicate involvement of bandgap states related to surface characteristics of the PNCs.

Compared to bulk or thin films, PQDs and PNCs tend to have a higher density of bandgap trap states that can substantially affect exciton or charge carrier dynamics. A typical signature of trap state involvement is shorter observed lifetime of the exciton and reduced PLQY. Trap states also tend to make the observed dynamics more complex, usually with multiple spectral bands and multi-exponential time decays, since the overall dynamics is usually not a simple first-order (single exponential) process [66].

One important factor to watch out in TA experiments is the pump power. Due to the very high peak power of fs laser pulses, it is very easy to generate multiple excitons or a high density of charge carriers in a small volume of space in the sample on a very short time scale, and this leads to nonlinear processes such as non-radiative exciton–exciton annihilation and Auger recombination that complicates the interpretation of the dynamics results. When possible, one should use low pump power to avoid such nonlinear processes. However, low pump power may not generate observable signal. Thus, one needs to optimize the pump power density by adjusting the laser spot size on the sample and to do careful control experiments always for each sample to determine the threshold for observing the nonlinear effect. For example, in Figure 3.42 above, pump power was varied to determine the threshold for nonlinear dynamics, which typically involves a new fast decay due to the nonlinear dynamics, as observed in many inorganic and organic semiconductor systems [74,75].

While fitting TA data with mathematical equations is relatively simple, detailed analysis to extract physical insight is not trivial. This is because there are more parameters involved than what experimental data can directly provide. With multiple experimental decays, as often observed, one usually assigns each exponential delay to a specific physical process. Strictly speaking, this is only approximate. To get deeper insight, we need physical models. From the physical model and with certain assumptions, one can generate computed data to compare to experimental data. One can do this iteratively by changing the physical model and assumptions made until the two sets of data fit. This will lead to a possible physical model with related time or rate constants [76]. Of course, like chemical reaction mechanisms in chemical kinetics, the model is usually only a possibility and is not unique. The normal practice is to select the simplest model possible first. One can in principle attempt to calculate the TA signal using theory. However, for complex materials like MHPs, approximations or assumption have to be made also in the computation. High level computations for TA are still expensive and often not feasible at the present time.

It is often challenging to determine which state corresponds with the observed lifetimes, especially in TA measurements. Time-resolved PL has more advantages since it is easier to identify the emitting excited state based on the PL spectrum.

It is often useful to decouple radiative and non-radiative processes to gain a better understanding of the exciton or charge carrier dynamics. There are two equations that relate the observed lifetime, τ_{ob}, to the radiative, τ_r, and non-radiative, τ_{nr}, lifetimes:

$$1/\tau_{ob} = 1/\tau_r + 1/\tau_{nr} \qquad (3.23)$$

The lifetimes are related to the PL quantum yield, Φ_{PL}, by

$$\Phi_{PL} = \tau_{ob}/\tau_r \qquad (3.24)$$

One should not confuse between radiative lifetime (τ_r) and observed lifetime (τ_{ob}), even when the latter is measured based on PL and often called *fluorescence lifetime*, which are only equal in the limit that τ_{nr} is very long or the PL quantum yield is near unity or 100%, such as in a perfect single crystal, according to Eqs. (3.23) and (3.24). Any lifetime measured experimentally, based on time resolved PL or TA measurements, is simply the observed lifetime τ_{ob}, which contains contributions from both τ_r and τ_{nr}. Non-radiative decays can be due to different mechanisms such as trapping due to bandgap states, charge transfer, or non-radiative energy transfer.

By experimentally measuring the observed lifetime and PLQY, one can use the two equations to extract the radiative and non-radiative lifetimes that can in turn provide deeper fundamental insights into the photophysical properties of the materials. It is important to note that the observed lifetime should be for the same excited state that produces the PL. For example, Figure 3.43 shows TRPL profile of three different sized $CH_3NH_3PbBr_3$ perovskite nanocrystals (PNCs) passivated using oleic acid and different amounts of (3-aminopropyl)triethoxysilane (APTES) [77]. The TRPL profiles are fit with a double exponential function with lifetimes that vary with the PNC size. Detailed analysis of the TRPL and PLQY leads to an understanding of how the radiative and non-radiative lifetimes depend on PNC size. It was found that the radiative lifetime is

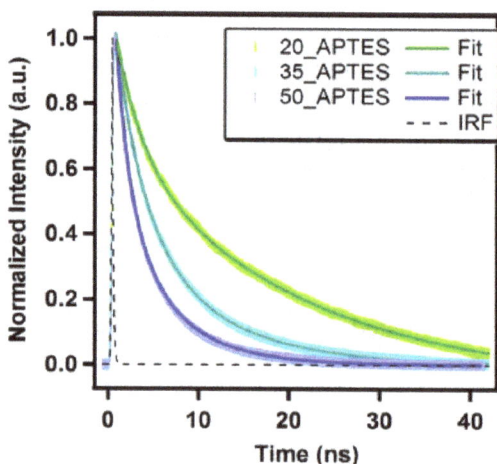

Figure 3.43.   TRPL decay traces of PNC films with different average PNC sizes (D) prepared with different amount of APTES with excitation (ex) and emission (em) wavelengths varied to match the peak excitonic absorption and PL positions: 20 μL (red, D = 9.3 nm, λ_{ex} = 520 nm, λ_{em} = 530 nm), 35 μL (green, D = 5.7 nm, λ_{ex} = 490 nm, λ_{em} = 500 nm), and 50 μL (blue, D = 3.1 nm, λ_{ex} = 440 nm, λ_{em} = 470 nm). Reproduced from Ref. [77].

more dominated by the PNC core, while the non-radiative lifetime is more dominated by surface characteristics determined by surface passivation with ligands.

3.8 Thermoelectrical Properties

Thermoelectrical materials are of strong interest for use in converting thermal energy into electricity. The thermal transport behavior of halide perovskites has received limited but increasing attention recently. Ulltralow thermal conductivity has been reported for $CH_3NH_3PbI_3$, $CsSnBr_{3-x}I_x$, and nanowires composed of $CsPbI_3$, $CsPbBr_3$, $CsSnI_3$. The conversion efficiency of a thermoelectric material is described by the dimensionless figure of merit $ZT = \alpha2\sigma T/\kappa$. An ideal thermoelectric material must possess a large Seebeck coefficient α, combined with a high electrical conductivity σ, and a low thermal conductivity κ. Most of the current state-of-the-art thermoelectric materials such as SnSe, PbTe, and Cu_2Se have low lattice thermal conductivity, similar in magnitude to that observed in MHPs. The high carrier mobility (50–$2,000$ cm^{2}V^{-1}s^{-1}) and large Seebeck coefficient (300–$5,500$ μVK^{-1}) claimed for both hybrid and all-inorganic halide perovskite single crystals, in conjunction with the ultralow thermal conductivity, make them promising as thermoelectrical materials.

For example, the thermal transport and thermoelectric properties of the $CsSnBr_{3-x}I_x$ perovskites have been studied very recently [78]. Figure 3.44 shows some of the experimentally determined thermoelectric properties of $CsSnBr_{3-x}I_x$($x = 0, 0.5, 1, 2$ and 3), including electrical conductivity, Seebeck coefficients, power factor, total thermal conductivity, lattice thermal conductivity, and figure of merit ZT [78]. A strong correlation between lattice dynamics and an ultralow thermal conductivity (reaching 0.32 Wm^{-1}K^{-1} at 550 K) has been found for a series $CsSnBr_{3-x}I_x$. The $CsSnBr_{3-x}I_x$ also exhibit reasonable Seebeck coefficient and controllable electrical transport properties. Experimental data and theoretical calculations suggest the Cs atom deviates from ideal cuboctahedral geometry imposed by the perovskite cage, behaving as a heavy atom rattler. This off-center feature, together with the distortion of SnX_6 octahedra, leads to a highly dynamic and disordered structure in $CsSnBr_{3-x}I_x$, and

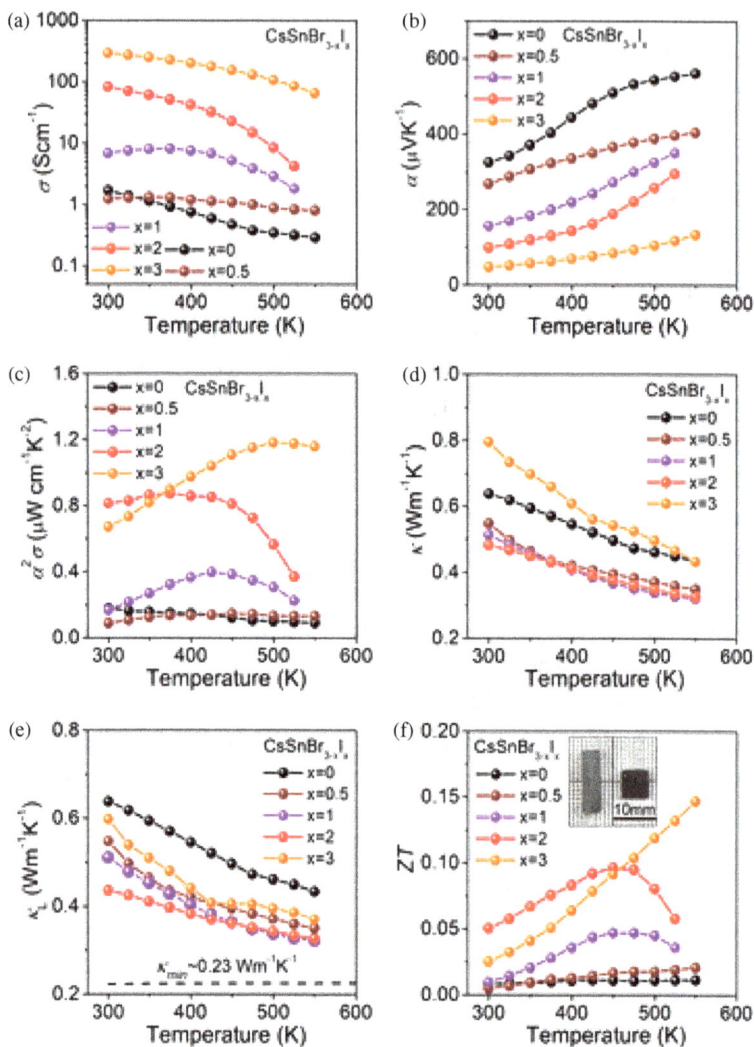

Figure 3.44. Thermoelectric properties of $CsSnBr_{3-x}I_x$ (x = 0, 0.5, 1, 2, and 3). (a) Electrical conductivity. (b) Seebeck coefficients. (c) Power factor. (d) Total thermal conductivity. (e) Lattice thermal conductivity, the dash line indicated the theoretical minimum lattice thermal conductivity of $CsSnBr_3$. (f) Figure of merit ZT; the inset shows the samples for the transport properties' measurements. Reproduced from ACS from Ref. [78].

thereby a very low Debye temperature and phonon velocity. In addition, there is indication of strong coupling between the low frequency optical phonons and heat carrying acoustical phonons, which results in strong phonon resonance scattering that induces the ultralow lattice thermal conductivity.

Doping is expected to affect the thermoelectrical properties. Study of doped MHPs for thermoelectrical applications has been very limited to date, and should be explored in the future. Effects of morphology and surface passivation, especially for nanostructured perovskites, could also be strong on thermoelectrical performance and will likely be the subject of future studies both experimentally and theoretically.

3.9 Theoretical and Computational Studies in Relation to Perovskite Properties

Besides extensive experimental studies, there have been a large number of theoretical and computational studies of MHPs in relation to their properties. Some examples have already been mentioned in previous sections. This section will provide a short and focused discussion on major computational techniques and their capabilities.

One of the most important and popular computational methods used to study MHPs is quantum mechanical density functional theory (DFT). It is used to predict crystal structures and stability and to calculate electronic band structures and energy levels, including bandgap states due to defects. One can also calculate related option properties such as absorption. Computation of lifetimes of charge carriers in the CB or defect states is more challenging and has only started to emerge recently. The calculated electronic band structure is useful in understanding and explaining experimental results such as electronic absorption and PL spectrum. One can also compute the vibrational spectrum or phonon modes to compare to Raman or IR results.

For instance, first-principles calculations based on non-collinear DFT have been used to study the atomic and electronic structures of a layered perovskite, $(C_6H_5C_2H_4NH_3)_2PbI_4$ [79]. The reduction of layer thickness between the bulk and a monolayer (ML) is investigated and compared to that of the benchmark OMHP, $(CH_3NH_3)PbI_3$, revealing that the bulkier $C_6H_5C_2H_4NH_3$ cations largely preserve the two-dimensional nature of the electronic structure in the layered bulk OMHP. Figure 3.45 shows the electronic band structures and PDOS of the PEA_2PbI_4 bulk and ML, respectively, as compared to that of $MAPbI_3$ bulk and MA_2PbI_4 ML. The states around the Fermi level mainly comprise p orbitals originating from the PbI_4 backbone. The valence band edge (VBE) states are of mainly iodine p character mixed with some Pb_s contributions. The CBE states, on the other hand, contain Pb_p orbitals. The latter are split into two separate contributions of p orbitals parallel to the PbI_2 plane (Pb p_x, p_y) and p_z orbitals perpendicular to this plane. Thus, the band gap is dominated by the PbI_2 layers. The differences between quasi-2D bulk and 2D ML are small, yielding almost the same gap as well as VB and CB regions. The PDOS of the PEA cations is narrower for the ML, due to absence of interactions to the next PEA sheet.

While DFT provides reliable structures and stabilities of perovskites, it often performs poorly in electronic structure predictions. The relativistic GW approximation has been demonstrated to be able to capture electronic structure accurately, but is computationally costly. Different approaches to improve the computation have been developed. For example, a recent study demonstrates an efficient and accurate bandgap calculation of MHPs by using the approximate quasiparticle DFT-1/2 method [80]. Using AMX_3 ($A = CH_3NH_3$, CH_2NHCH_2, Cs; M = Pb, Sn, X = I, Br, Cl) as demonstration, the influence of the crystal structure, variation of ions, and relativistic effects on the electronic structures are systematically studied and compared with experimental results, as shown in Figure 3.46. The results show that the DFT-1/2 method produces accurate bandgaps with the precision of the GW method with no more computational cost than standard DFT. This enables accurate electronic structure prediction of complex halide perovskite structures and design for new lead-free materials.

High-throughput computational design has been applied to study MHPs. For example, it has been used to design lead-free hybrid halide semiconductors with robust stability and desired properties [81]. On the basis of 24 prototype structures that include perovskite and non-perovskite structures and several typical organic cations, a comprehensive quantum materials' repository that contains 4,507 hypothetical hybrid compounds is built using large-scale first-principles calculations. After a high-throughput screening of this repository, 23 candidates are identified

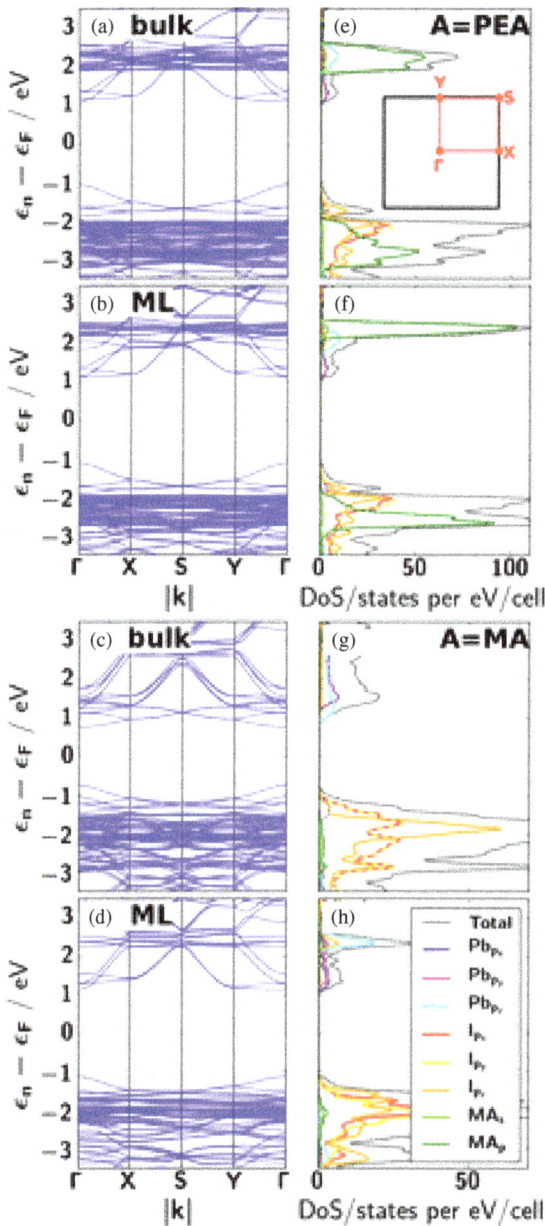

Figure 3.45. Scalar-relativistic band structures (a–d) and projected density of states (PDOS) (e–h) of PEA$_2$PbI$_4$ bulk (a, e), monolayer (ML) (b, f), MAPbI$_3$ bulk (c, g), and MA$_2$PbI$_4$ ML(d, h), respectively. The inset in panel *e* shows the BZ and the sampled path along high symmetry points. Reproduced from Ref. [79].

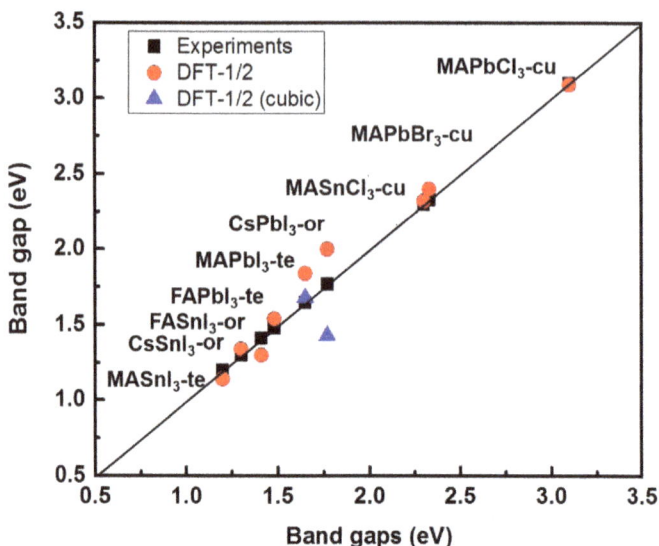

Figure 3.46. Comparison between experimental and DFT-1/2 bandgaps of AMX$_3$ perovskites. Note: when the room-temperature crystal structure is uncertain (CH$_3$NH$_3$PbI$_3$ and CSPbI$_3$), the band gaps of the high-temperature crystal structure, i.e., the cubic phase, are also plotted. When no experimental result is available (CH$_3$NH$_3$SnCl$_3$), the GW result is used. MA = CH$_3$NH$_3$; FA = CH$_2$NHCH$_2$. Reproduced from Ref. [80].

for LEDs and 13 candidates for solar energy conversion. This example demonstrates a new approach to designing novel organic–inorganic functional materials by exploring a great variety of prototype structures.

One interesting recent development in predicting or understanding the properties of perovskites uses statistical tools based on machine learning (ML) and deep networks. For example, ML has been utilized to predict the structural and electronic properties of MHPs using a hierarchical convolutional neural network [82]. Convolutional neural networks have been used to develop a predictive model for the electronic properties of MHPs with a billions-range materials design space. In this architecture, each neural network element has a designated role in the estimation process from predicting complex features of the perovskites, such as lattice constant and octahedral till angle, to narrowing down possible ranges for the values of interest. The study illustrates the importance of a careful network design and hierarchical approach to alleviating problems associated with imbalanced dataset distributions, which is common in materials dataset.

In a recent review [83], an overview was provided on the state of the art in the automated MHP synthesis and existing methods for navigating multicomponent compositional space. It highlights limitations and pitfalls of the existing strategies and formulates the requirements for necessary ML tools including causal and Bayesian methods, as well as strategies based on co-navigation of theoretical and experimental spaces. We argue that ultimately the goal of automated experiments is to simultaneously optimize the materials synthesis and refine the theoretical models that underpin target functionalities. Furthermore, the near-term development of automated experimentation will not lead to the full exclusion of the human operator but rather automatization of repetitive operations, deferring human role to high-level slow decisions. We also discuss the emerging opportunities leveraging ML-guided automated synthesis to the development of high-performance perovskite optoelectronics. Figure 3.47 shows a comparison between classical ML and casual ML that differs in assumption about relationships between observed and non-observed variables [8].

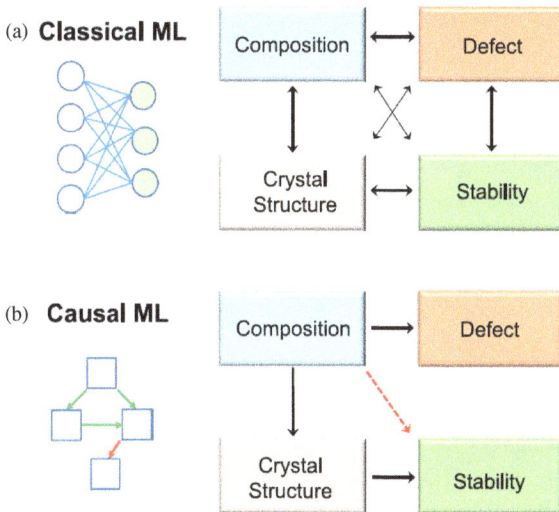

Figure 3.47. Classical versus causal machine learning (ML). (a) Compared with physics learning, in (A) classical ML the correlative relationships between the variables are established and the causal links are ignored, often leading to paradoxical results. (b) In causal ML, some (or none) of the relationships between observed and non-observed variables are assumed to be known, and discovery of the presence and functional form of the others is the goal of the analysis. Reproduced from Ref. [83].

3.10 Summary

This chapter covers several important aspects of the properties and functionalities of MHPs, including structural, electronic, optical, magnetic, and dynamic. Several different types of perovskites are discussed and compared, including the common ABX_3 type, double perovskites, 2D perovskites, and various doped or alloyed perovskites. The examples presented illustrate the diversity of structures and properties. With further research, one can anticipate new structures and properties to be discovered.

References

[1] Z.P. Lian, Q.F. Yan, Q.R. Lv, Y. Wang, L.L. Liu, L.J. Zhang, S.L. Pan, Q. Li, L.D. Wang, and J.L. Sun, *Sci. Rep.* 5, 10 (2015).

[2] Y.Y. Dang, Y. Liu, Y.X. Sun, D.S. Yuan, X.L. Liu, W.Q. Lu, G.F. Liu, H.B. Xia, and X.T. Tao, *Crystengcomm.* 17 (3), 665 (2015).

[3] T. Oku, in *Solar Cells — New Approaches and Reviews*, edited by Leonid A. Kosyachenko (Intech Open, 2015), p. 77.

[4] J. Calabrese, N.L. Jones, R.L. Harlow, N. Herron, D.L. Thorn, and Y. Wang, *J. Am. Chem. Soc.* 113 (6), 2328 (1991).

[5] P.S. Whitfield, N. Herron, W.E. Guise, K. Page, Y.Q. Cheng, I. Milas, and M.K. Crawford, *Sci. Rep.* 6 (2016).

[6] M.A. Perez-Osorio, R.L. Milot, M.R. Filip, J.B. Patel, L.M. Herz, M.B. Johnston, and F. Giustino, *J. Phys. Chem.* C 119 (46), 25703 (2015).

[7] C. Quarti, G. Grancini, E. Mosconi, P. Bruno, J.M. Ball, M.M. Lee, H.J. Snaith, A. Petrozza, and F. De Angelis, *J. Phys. Chem. Lett.* 5 (2), 279 (2014).

[8] M.L. Liao, B.B. Shan, and M. Li, *J. Phys. Chem. Lett.* 10 (6), 1217 (2019).

[9] K. Xu, E.T. Vickers, L.S. Rao, S.A. Lindley, A.C. Allen, B.B. Luo, X.M. Li, and J.Z. Zhang, *Chem.-Eur. J.* 25 (19), 5014 (2019).

[10] C.C. Stoumpos, C.D. Malliakas, and M.G. Kanatzidis, *Inorg. Chem.* 52 (15), 9019 (2013).

[11] E.M. Tennyson, T.A.S. Doherty, and S.D. Stranks, *Nat. Rev. Mater.* 4 (9), 573 (2019).

[12] J. Yang, C. Lei, C. Rong, and K.M. Jiang, *Eur. Phys. J.* B 62 (3), 263 (2008).

[13] R. Cheng, M.W. Daniels, J.-G. Zhu, and D. Xiao, *Sci. Rep.* 6, 24223 (2016).

[14] G. Long, C. Jiang, R. Sabatini, Z. Yang, M. Wei, L.N. Quan, Q. Liang, A. Rasmita, M. Askerka, G. Walters, X. Gong, J. Xing, X. Wen, R. Quintero-Bermudez, H. Yuan, G. Xing, X.R. Wang, D. Song, O. Voznyy, M. Zhang, S. Hoogland, W. Gao, Q. Xiong, and E.H. Sargent, *Nat. Photo.* 12 (9), 528 (2018).

[15] C. Lan, Z. Zhou, R. Wei, and J.C. Ho, *Mater. Today Energy* 11, 61 (2019).

[16] B.A. Connor, L. Leppert, M.D. Smith, J.B. Neaton, and H.I. Karunadasa, *J. Am. Chem. Soc.* 140, 5235 (2018).

[17] M.S. Lassoued, L.-Y. Bi, Z. Wu, G. Zhou, and Y.-Z. Zheng, *J. Mater. Chem. C* 8 (16), 5349 (2020).

[18] H. Tang, Y. Xu, X. Hu, Q. Hu, T. Chen, W. Jiang, L. Wang, and W. Jiang, *Adv. Sci.* 8, 2004118 (2021).

[19] Y. Zhou, J. Chen, O.M. Bakr, and H.-T. Sun, *Chem. Mat.* 30 (19), 6589 (2018).

[20] M. Cong, B. Yang, F. Hong, T. Zheng, Y. Sang, J. Guo, S. Yang, and K. Han, *Sci. Bull.* 65 (13), 1078 (2020).

[21] K. Zheng and T. Pullerits, *J. Phys. Chem. Lett.* 10 (19), 5881 (2019).

[22] C. Ge, Y.Z.B. Xue, L. Li, B. Tang, and H. Hu, *Front. Mater.* 7 (2020).

[23] T. Zhu, Y. Yang, K. Gu, C. Liu, J. Zheng, and X. Gong, *ACS Appl. Mater. Interfaces* 12 (46), 51744 (2020).

[24] S.D. Stranks, G.E. Eperon, G. Grancini, C. Menelaou, M.J.P. Alcocer, T. Leijtens, L.M. Herz, A. Petrozza, and H.J. Snaith, *Science* 342 (6156), 341 (2013).

[25] M. Gratzel, *Nat. Mater.* 13 (9), 838 (2014).

[26] T. Leijtens, T. Giovenzana, S.N. Habisreutinger, J.S. Tinkham, N.K. Noel, B.A. Kamino, G. Sadoughi, A. Sellinger, and H.J. Snaith, *ACS Appl. Mater. Interfaces* 8 (9), 5981 (2016).

[27] W. Tress, K. Domanski, B. Carlsen, A. Agarwalla, E.A. Alharbi, M. Graetzel, and A. Hagfeldt, *Nat. Energy* 4 (7), 568 (2019).

[28] M. Green, E. Dunlop, J. Hohl-Ebinger, M. Yoshita, N. Kopidakis, and X. Hao, *Prog. Photovolt.* 29 (1), 3 (2021).

[29] J. Wang, C. Zhang, H. Liu, R. McLaughlin, Y. Zhai, S.R. Vardeny, X. Liu, S. McGill, D. Semenov, H. Guo, R. Tsuchikawa, V.V. Deshpande, D. Sun, and Z.V. Vardeny, *Nat. Commun.* 10 (2019).

[30] Y. Zhai, S. Baniya, C. Zhang, J. Li, P. Haney, C.-X. Sheng, E. Ehrenfreund, and Z.V. Vardeny, *Sci. Adv.* 3 (7) (2017).

[31] E. Meyer, D. Mutukwa, N. Zingwe, and R. Taziwa, *Metals* 8 (9) (2018).

[32] A.a.O. El-Ballouli, O.M. Bakr, and O.F. Mohammed, *J. Phys. Chem. Lett.* 11 (14), 5705 (2020).

[33] E.-B. Kim, M.S. Akhtar, H.-S. Shin, S. Ameen, and M.K. Nazeeruddin, *J. Photochem. Photobiol. C: Photochem. Rev.* 48 (2021).

[34] X. Li, B. Traore, M. Kepenekian, L. Li, C.C. Stoumpos, P. Guo, J. Even, C. Katan, and M.G. Kanatzidis, *Chem. Mat.* 33 (15), 6206 (2021).

[35] Y. Yao, B. Kou, Y. Peng, Z. Wu, L. Li, S. Wang, X. Zhang, X. Liu, and J. Luo, *Chem Commun.* 56 (21), 3206 (2020).

[36] D. Li, X. Liu, W. Wu, Y. Peng, S. Zhao, L. Li, M. Hong, and J. Luo, *Angew. Chem.-Int. Edit.* 60 (15), 8415 (2021).

[37] N.K. Nandha and A. Nag, *Chem Commun.* 54 (41), 5205 (2018).

[38] W. Ke and M.G. Kanatzidis, *Nat. Commun.* 10 (2019).

[39] T.L. Barr, *Modern ESCA: The Principles and Practice of X-Ray Photoelectron Spectroscopy* (CRC Press, Boca Raton, FL, 2008), p. 376.

[40] Z. Ahmad, M.A. Najeeb, R.A. Shakoor, A. Alashraf, S.A. Al-Muhtaseb, A. Soliman, and M.K. Nazeeruddin, *Sci. Rep.* 7, 15406 (2017).

[41] D. Niesner, *APL Mater.* 8 (9), 21 (2020).

[42] F.S. Zu, P. Amsalem, D.A. Egger, R.B. Wang, C.M. Wolff, H.H. Fang, M.A. Loi, D. Neher, L. Kronik, S. Duhm, and N. Koch, *J. Phys. Chem. Lett.* 10 (3), 601 (2019).

[43] A.M.A. Leguy, P. Azarhoosh, M.I. Alonso, M. Campoy-Quiles, O.J. Weber, J.Z. Yao, D. Bryant, M.T. Weller, J. Nelson, A. Walsh, M. van Schilfgaarde, and P.R.F. Barnes, *Nanoscale* 8 (12), 6317 (2016).

[44] L.M. Herz, in *Annual Review of Physical Chemistry, Vol 67*, edited by M. A. Johnson and T. J. Martinez (Annual Reviews, Palo Alto, 2016), Vol. 67, p. 65.

[45] L.M. Herz, *ACS Energy Lett.* 2 (7), 1539 (2017).

[46] Y. Chen, H.T. Yi, X. Wu, R. Haroldson, Y.N. Gartstein, Y.I. Rodionov, K.S. Tikhonov, A. Zakhidov, X.Y. Zhu, and V. Podzorov, *Nat. Commun.* 7, 9 (2016).

[47] J.I. Pankove, *Optical Properties in Semiconductors* (Dover Publications, Inc., New York, 1971), p. 422.

[48] J. Bardeen, F.J. Blatt, and L.H. Hall, presented at the Atlantic City Photoconductivity Conference, Atlantic City, 1954 (pub 1956) (unpublished).

[49] D.Q. Chen, X. Chen, Z.Y. Wan, and G.L. Fang, *ACS Appl. Mater. Interfaces* 9 (24), 20671 (2017).

[50] R.Z. Li, C. Yi, R. Ge, W. Zou, L. Cheng, N.N. Wang, J.P. Wang, and W. Huang, *Appl. Phys. Lett.* 109 (15), 4 (2016).

[51] P.G. Han, X. Zhang, C. Luo, W. Zhou, S.Q. Yang, J.Z. Zhao, W.Q. Deng, and K.L. Han, *ACS Central Sci.* 6 (4), 566 (2020).

[52] C. Yuan, X. Li, S. Semin, Y. Feng, T. Rasing, and J. Xu, *Nano Lett.* 18 (9), 5411 (2018).

[53] J.L. Xu, X.Y. Li, J.B. Xiong, C.Q. Yuan, S. Semin, T. Rasing, and X.H. Bu, *Adv. Mater.* 32 (3), 13 (2020).

[54] P. Brenner, O. Bar-On, M. Jakoby, I. Allegro, B.S. Richards, U.W. Paetzold, I.A. Howard, J. Scheuer, and U. Lemmer, *Nat. Commun.* 10, 7 (2019).

[55] S. Sil, H. Luitel, J. Dhar, M. Chakrabarti, P.P. Ray, B. Bandyopadhyay, and D. Sanyal, *Phys. Lett. A* 384 (14) (2020).

[56] B. Nafradi, P. Szirmai, M. Spina, H. Lee, O.V. Yazyev, A. Arakcheeva, D. Chernyshov, M. Gibert, L. Forro, and E. Horvath, *Nat. Commun.* 7, 8 (2016).

[57] L. Protesescu, S. Yakunin, M.I. Bodnarchuk, F. Krieg, R. Caputo, C.H. Hendon, R.X. Yang, A. Walsh, and M.V. Kovalenko, *Nano Lett.* 15 (6), 3692 (2015).

[58] F. Zhang, H. Zhong, C. Chen, X.-g. Wu, X. Hu, H. Huang, J. Han, B. Zou, and Y. Dong, *ACS Nano* 9 (4), 4533 (2015).

[59] T.C. Jellicoe, J.M. Richter, H.F.J. Glass, M. Tabachnyk, R. Brady, S.E. Dutton, A. Rao, R.H. Friend, D. Credgington, N.C. Greenham, and M.L. Bohm, *J. Am. Chem. Soc.* 138 (9), 2941 (2016).

[60] W.Y. Liu, Q.L. Lin, H.B. Li, K.F. Wu, I. Robel, J.M. Pietryga, and V.I. Klimov, *J. Am. Chem. Soc.* 138 (45), 14954 (2016).

[61] A. Kausar, A. Sattar, C. Xu, S. Zhang, Z. Kang, and Y. Zhang, *Chem. Soc. Rev.* 50 (4), 2696 (2021).

[62] C. Liu, J. Sun, W.L. Tan, J. Lu, T.R. Gengenbach, C.R. McNeill, Z. Ge, Y.-B. Cheng, and U. Bach, *Nano Lett.* 20 (2), 1240 (2020).

[63] D.K. Sharma, S. Hirata, and M. Vacha, *Nat. Commun.* 10, 9 (2019).

[64] Y.A. Darmawan, M. Yamauchi, and S. Masuo, *J. Phys. Chem. C* 124 (34), 18770 (2020).

[65] J.Z. Zhang, *Optical Properties and Spectroscopy of Nanomaterials* (World Scientific Publisher, Singapore, 2009), p. 383.

[66] D. Wheeler and J.Z. Zhang, *Adv. Mater.* 25, 2878 (2013).

[67] J. Even, L. Pedesseau, and C. Katan, *J. Phys. Chem. C* 118 (22), 11566 (2014).

[68] M.A. Becker, R. Vaxenburg, G. Nedelcu, P.C. Sercel, A. Shabaev, M.J. Mehl, J.G. Michopoulos, S.G. Lambrakos, N. Bernstein, J.L. Lyons, T. Stoferle, R.F. Mahrt, M.V. Kovalenko, D.J. Norris, G. Raino, and A.L. Efros, *Nature* 553 (7687), 189 (2018).

[69] D. Giovanni, H. Ma, J. Chua, M. Graetzel, R. Ramesh, S. Mhaisalkar, N. Mathews, and T.C. Sum, *Nano Lett.* 15 (3), 1553 (2015).

[70] D.V. Bulaev and D. Loss, *Phys. Rev. B* 71 (20) (2005).

[71] M. Kepenekian and J. Even, *J. Phys. Chem. Lett.* 8 (14), 3362 (2017).

[72] J. Li and P.M. Haney, *Phys. Rev. B* 93 (15) (2016).

[73] B. Luo, Y.C. Pu, Y. Yang, S.A. Lindley, G. Abdelmaged, H. Ashry, Y. Li, X.M. Li, and J.Z. Zhang, *J. Phys. Chem. C* 119 (47), 26672 (2015).

[74] T.W. Roberti, N.J. Cherepy, and J.Z. Zhang, *J. Chem. Phys.* 108 (5), 2143 (1998).

[75] M. Kreger, N. Cherepy, J. Zhang, J. Scott, G. Klaerner, R. Miller, D. McBranch, B. Kraabel, and S. Xu, *Phys. Rev. B* 61 (12), 8172 (2000).

[76] Y.C. Pu, M.G. Kibria, Z.T. Mi, and J.Z. Zhang, *J. Phys. Chem. Lett.* 6 (13), 2649 (2015).

[77] S.B. Naghadeh, B.B. Luo, Y.C. Pu, Z. Schwartz, W.R. Hollingsworth, S.A. Lindley, A.S. Brewer, A.L. Ayzner, and J.Z. Zhang, *J. Phys. Chem. C* 123 (7), 4610 (2019).

[78] H.Y. Xie, S.Q. Hao, J.K. Bao, T.J. Slade, G.J. Snyder, C. Wolverton, and M.G. Kanatzidis, *J. Am. Chem. Soc.* 142 (20), 9553 (2020).

[79] J. Gebhardt, Y. Kim, and A.M. Rappe, *J. Phys. Chem. C* 121 (12), 6569 (2017).

[80] S.X. Tao, X. Cao, and P.A. Bobbert, *Sci. Rep.* 7, 9 (2017).

[81] Y.H. Li and K.S. Yang, *Energy Environ. Sci.* 12 (7), 2233 (2019).

[82] W.A. Saidi, W. Shadid, and I.E. Castelli, *Npj Comput. Mater.* 6 (1), 7 (2020).

[83] M. Ahmadi, M. Ziatdinov, Y.Y. Zhou, E.A. Lass, and S.V. Kalinin, *Joule* 5 (11), 2797 (2021).

Chapter 4

Applications of Metal Halide Perovskites

Metal halide perovskites (MHPs) have attracted extensive attention for different technological applications such as solar cells, photodetectors, light-emitting diodes (LEDs), lasers, sensors, and photocatalysis due to their excellent optoelectronic properties including large absorption cross-section, high photoluminescence (PL) quantum yield (QY), and long carrier diffusion length, etc. This chapter covers typical examples on the emerging device applications of MHPs with focus on the basic operating principles, requirements of materials properties, and key performance measurements.

4.1 Photovoltaics

The first- and second-generation solar cells, i.e., silicon wafer-based and thin-film solar cells have performed well in terms of efficiency and stability. However, ultra-high purity silicon (>99.999%) with the high cost of materials and processing of the wafers have prevented the widespread use of such solar cells as alternatives to fossil-fuel-based energy sources such as thermal power generation (present cost <0.05USD per kWh of electric power). MHPs were originally considered as a curious replacement to dye molecules in mesoscopic sensitized solar cells. However, it was quickly realized that MHPs are unique semiconductor materials for photovoltaic (PV) applications, different from dye molecules and other organic absorbers. Although organic solar cells and dye-sensitized solar cells (DSSCs) could be processed relatively cheaply, their low long-term stability has limited their commercialization. Since the first embodiment of perovskites solar cells (PSCs) showing a photocurrent with a power conversion efficiency (PCE) of around 3–4%, the PCE values have increased to over 25% within a few years [1]. The combination of high PCE and low-cost of MHPs makes PSCs promising for PV applications. The operating principles and key parameters of representative PSCs are discussed. In addition, a series of single-junction or tandem solar cells, including organic–inorganic halide perovskites, all-inorganic perovskites, and lead-free and inorganic perovskites, are compared.

4.1.1 *Device Structure of PSCs*

For PSCs, both preparation method and device architecture are key factors. PSCs can be generally classified into four different configurations: meso-scopic-adsorbed structure, normal/inverted planar structure, and mesoscopic-filled structure, as shown in Figure 4.1 [2]. Typically, PSCs have layered structures comprising a transparent conductive oxide (TCO)-coated glass substrate, an n-type semiconductor as the electron-transport layer (ETL), a perovskite absorber layer, a p-type semiconductor as the hole-transport layer (HTL), and a back-contact (metal, TCO, or carbon). For mesoscopic PSCs, the perovskite absorber sensitizes a mesoporous metal oxide layer used as a scaffold, e.g., meso-TiO_2. In planar PSCs, the perovskite absorber layer is sandwiched between a planar ETL, e.g., compact TiO_2 (c-TiO_2), SnO_2, or C_{60} and its derivatives, and an HTL, e.g., spiro-OMeTAD, poly(triarylamine) (PTAA), and poly(3,4-ethylenedioxythiophene)-poly(styrene-sulfonate) (PEDOT:PSS), to realize the separation of electrons and holes. Depending on the sequence of depositing the ETL and HTL and different directions of carrier transmission, the structure can be divided into two categories: formal (conventional, n-i-p) versus inverted (n-i-p) architectures. Notably, the design and fabrication of inverted devices is not a simple reversal of the formal device structure, but the two device structures typically require systematic development of suitable ETL and HTL, especially in light of the restrictions in the suitability of processing solvents. In triple mesoscopic

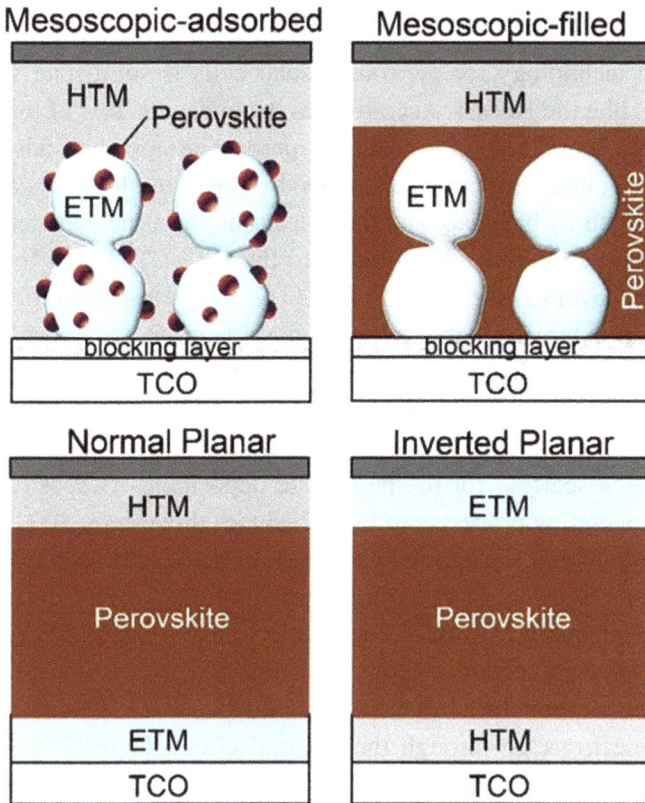

Figure 4.1. Four device configurations of PSCs: mesoscopic-adsorbed structure, normal/inverted planar structure, and mesoscopic-filled structure. Reproduced from Ref. [2].

PSCs, the perovskite is deposited on a triple-layer scaffold, covering a screen-printed mesoporous TiO_2 layer, a ZrO_2 spacer layer, and a carbon electrode. Such devices use carbon material as electrodes to replace the noble metal back-contacts and do not require a hole-conducting layer. Compared with single-junction devices, multi-junction (tandem) solar cells (TSCs) consisting of multiple light absorbers with considerably different bandgaps show great potential in breaking the Shockley–Queisser (S–Q) efficiency limit (30%) of a single junction solar cell by absorbing light in a broader range of wavelengths. The performance of PSCs depends not only on their architecture, but also heavily on the type of perovskite material used.

4.1.2 *The Operating Principle of PSCs*

The basic technology for perovskite solar cells is solid-state sensitized solar cells like the DSSCs. According to the understanding of the work of mesoscopic sensitized solar cells, the perovskite absorber, adsorbed on TiO$_2$ surface, undergoes photoexcitation by absorbing light of wavelength matching with its bandgap, and then electrons from the conduction band (CB) of the perovskite are injected into the CB of TiO$_2$ and transported through the layer to reach the TCO (FTO or ITO) substrate. From there, it flows in an external circuit to the counter electrode (cathode), where it combines with oxidized species of electrolyte. The mesoporous (TiO$_2$ or Al$_2$O$_3$) layer serves to drive the photoinjected electrons to the blocking layer, but it is not essential for operation of the PSCs since its main role is to serve as a scaffold for the perovskite deposition. As sensitized solar cells, the operating processes of PSCs includes three steps: (i) charge carrier generation and separation by incident light, (ii) carrier transport, and (iii) carrier collection. In the typical configuration of a PSC device, as the perovskite absorbs light, an electron–hole pair is created as per Eq. (4.1) and then photogenerated electrons are injected into ETL (mesoporous semiconductor) and the holes are driven into the HTL as per Eq. (4.2). The injected electrons go through the external circuit to reach the cathode while the holes are transported to the metal anode [3]. At the interface of HTL/back contact, the hole and electron recombine regenerating the system. The entire operation process of PSCs is displayed in Figure 4.2. The p-n junctions are responsible for the creation of a built-in electric field that allows the charge separation, where electrons move to the mesoporous materials (e.g., TiO$_2$, Al$_2$O$_3$, and SnO$_2$) and holes move to the HTL (spiro-OMeTAD, PTAA, etc.) [4]. Since both materials have different Fermi levels, charges flow until equilibrium is reached; a space-charge region appears at the respective interface. In this regard, the summary in Figure 4.2 is comparison of consequent band bending.

$$Perovskite + hv \rightarrow e^-...h^+)_{Perovskite} \tag{4.1}$$

$$(e^-...h^+)_{Perovskite} \rightarrow e^-(TiO_2/ETL) + h^+(HTL) \tag{4.2}$$

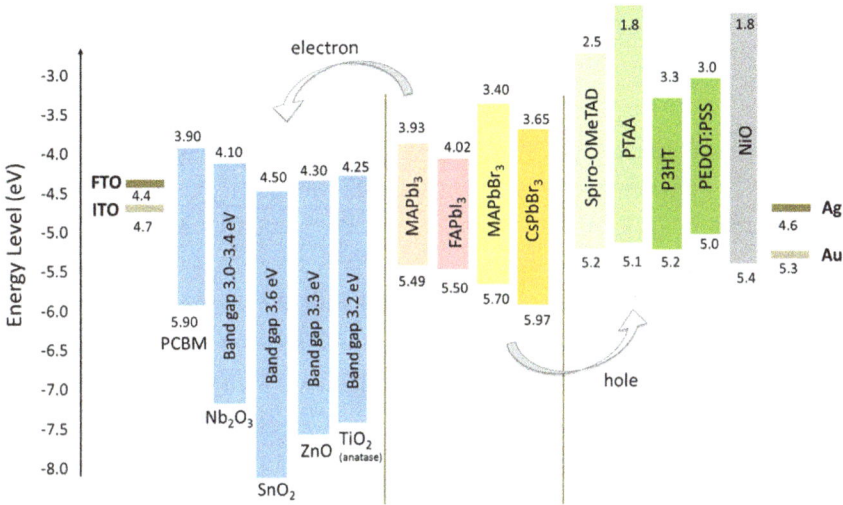

Figure 4.2. Energy levels of lead halide perovskite absorbers and various metal oxide electron transport materials and hole transport materials (HTMs) employed in solar cell devices. In the perovskites, MA and FA stand for methylammonium and formamidinium, respectively. PCBM denotes [6,6]-phenyl-C 61-butyric acid methyl ester. In HTMs, PTAA, P3HT, and PEDOT:PSS denote poly[bis(4-phenyl)(2,4,6-trimethylphenyl)amine, poly(3-hexylthiophene-2,5-diyl), and poly(3,4-ethylenedioxythiophene)-polystyrene sulfonate, respectively. Reproduced from Ref. [4].

However, competing with the extraction of photogenerated charges, undesired reactions may occur. The main back reactions involve: (i) exciton decay by photoluminescence (Eq. (4.3)), (ii) non-radiative recombination (Eq. (4.4)), and (iii) recombination of the charge carriers at the interfaces (TiO$_2$ surface, HTL surface, and TiO$_2$/HTL interface) with consequent heat release (Eqs. (4.5)–(4.7)).

$$(e^-...h^+)_{Perovskite} \rightarrow hv' \tag{4.3}$$

$$(e^-...h^+)_{Perovskite} \rightarrow \nabla \tag{4.4}$$

$$e^-(TiO_2) + h^+(Perovskite) \rightarrow \nabla \tag{4.5}$$

$$e^-(Perovskite) + h^+(HTL) \to \nabla \qquad (4.6)$$

$$e^-(TiO_2) + h^+(HTL) \to \nabla \qquad (4.7)$$

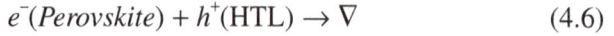

In general, the key parameters for evaluating the performance of solar cells can be summarized as: (1) open-circuit voltage (V_{oc}), (2) short-circuit current density (J_{sc}), (3) fill factor (FF), (4) PCE, and (5) external quantum efficiency (EQE).

Under light, an ideal solar cell can be equivalent to a parallel circuit of a diode and a current source. According to the relationship between the port current and voltage of this circuit, we can get:

$$I = I_D - I_{ph} = I_0 \left[\exp\left(\frac{qv}{nk_BT} \right) - 1 \right] - I_L \qquad (4.8)$$

where I_D and I_{ph} are the intensity of dark current and photocurrent, respectively, I_0 is the diode saturation current, q is the electron charge (C), k_B is Boltzmann's constant (J/K), n is the ideal factor, and T is the ambient temperature (K). The I–V curve of a solar cell is shown in Figure 4.3. The maximum current value at ordinate is the short-circuit current I_{sc}, and the maximum voltage value at abscissa is the open circuit voltage V_{oc}.

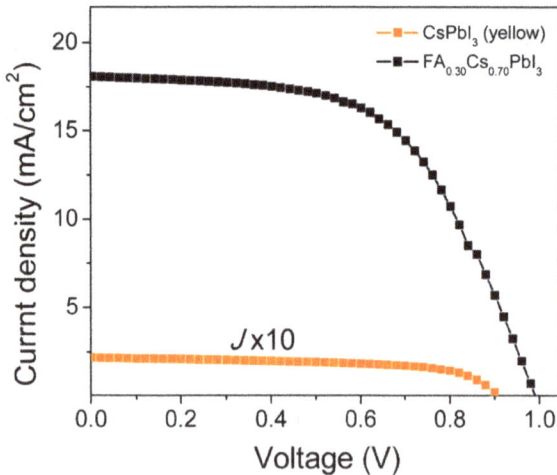

Figure 4.3.   The I–V curve of a typical solar cell. Reproduced from Ref. [5].

Open-circuit voltage (V_{oc}) is the voltage when the electrodes at both terminals of the solar cell are disconnected in the light. It is also the maximum voltage that the solar cell can output. V_{oc} can be derived based on Eq. (4.8) ($I = 0$), as follows:

$$V_{oc} = \left(\frac{nKT}{q}\right)\ln\left(\frac{I_{ph}}{I_0} + 1\right) \tag{4.9}$$

The value of V_{oc} is related to the reverse saturation current I_0. And I_0 is determined by the various recombination mechanisms of solar cells, so usually the V_{oc} can be used to detect the recombination of carriers in solar cells. In addition, the open circuit voltage V_{oc} in the solar cell is associated with the quasi-Fermi level of the transport layer and the perovskite light-absorbing layer in the semiconductor.

Short-circuit current density (J_{sc}) is the output current density of the battery when the two terminals of the solar cell are short-circuited in the light. The short-circuit current density is equal to the ratio of the photocurrent intensity to the effective area of the solar cell. The *FF* is related to the maximum power output of the solar cell, with the following associated expression:

$$FF = \frac{V_{max}J_{max}}{V_{oc}J_{sc}} \times 100\% \tag{4.10}$$

V_{max} and J_{max} are the voltage and current density values corresponding to the maximum output power (P_{max}), respectively. The *FF* value is closely related to the performance of solar cells. Generally, the *FF* of solar cells with better performance can reach more than 80%. The smaller the series resistance of the solar cell and the larger the parallel resistance, the larger the *FF*. The current–voltage characteristic curve reflected in the solar cell is that the closer the curve is to a square, the larger the fill factor.

PCE, also known as the photoelectric conversion efficiency, is the ratio of the maximum output power (P_{max}) of the solar cell to the input power (P_{in}) of the incident sunlight. PCE is defined as Eq. (4.11):

$$PCE = \frac{P_{max}}{P_{in}} = \frac{V_{max}J_{max}}{P_{in}} = \frac{FFV_{oc}J_{sc}}{P_{in}} \times 100\% \tag{4.11}$$

PCE is the most important parameter reflecting the performance of solar cells. The value of one is closely relevant to the spectral distribution of incident light, energy intensity, and ambient temperature. In order to make the test data comparable, the PCE measurement is generally carried out under AM 1.5G 100 mW cm^{-2} light and 25 °C. Based on the above formula, to maximize the efficiency of solar cells, FF, I_{sc} and V_{oc} should all be maximized.

EQE is defined as the ratio of the number of electrons (N_e) produced by the solar cell per unit time to the number of photons of incident monochromatic light (N_p), which can be described as

$$EQE(\lambda) = \frac{N_e}{N_p} = \frac{1,240}{\lambda\,(nm)} \times \frac{J_{sc}(\lambda)}{P_{in}(\lambda)} \times 100\% \qquad (4.12)$$

where λ is the incident wavelength and P_{in} is the incident light power (mA cm^{-2}).

4.1.3 *Advantages of PSCs*

As absorber material of PSC devices, the perovskite family typically used is based on organic–inorganic lead perovskites with the polycrystalline structure $CH_3NH_3PbX_3$, where X is a halide atom (I, Cl, Br, or a mixture of them). This type of materials shows advantageous properties to be used as a PV absorber, i.e., (1) strong optical absorption due to s-p antibonding coupling, (2) high electron and hole mobilities and diffusion lengths, (3) superior structural defect tolerance and shallow point defects, (4) low surface recombination rate, and (5) favorable grain boundaries (GBs) preventing the electron–hole recombination. In fact, compared to the most common PV systems (Table 4.1), perovskite semiconductors show tunable bandgap, balanced electron/hole transportation, low recombination rate, carrier lifetime over 100 ns, and a diffusion length over 1,000 nm. These superior PV properties are originated from the combination of high perovskite symmetry (Oh), high electronic dimensionality, the unique atomic electronic configuration of Pb including the lone-pair Pb 6s, the inactive Pb 6p, the strong spin-orbit coupling (SOC) along with the arrangement of polar organic cations, and the ionic nature of halides.

Table 4.1. Physical properties of most common photovoltaic materials.

	Perovskite	Si	CIGS	GaAs	CdTe
Band gap/eV	1.5 (tunable)	1.1	1.12	1.43	1.5
Absorption coefficient/cm^{-1}	10^{4-5}	10^3	10^{4-5}	10^{4-5}	10^3
Carrier concentration/cm^{-3}	10^{16-17}	10^{16}	10^{15-16}	10^{17}	10^{14-15}
Carrier mobility/cm^{2}V^{-1}s^{-1}	Up to 2,000	1,500	$\leq$10	8,500	10
Carrier lifetime	$\geq$100 ns	ms	50–200 ns	$\leq$100 ns	20 ns

High optical absorption is very desirable for realizing efficient and low-cost thin-film solar cells. Lead halide perovskites, such as methyl ammonium lead iodide ($CH_3NH_3PbI_3$ or $MAPbI_3$), exhibit extremely high optical absorption coefficients. The Pb-based PSCs with a thinner absorber layer thickness of $\approx$500 nm accomplish record and similar PCEs in comparison with mainstream inorganic thin-film solar cell absorber layers such as GaAs, Cu(In,Ga)Se$_2$ (CIGS), and CdTe. The optical absorption of a semiconductor at photonic energy $\hbar\omega$ is directly obtained by

$$\frac{2\pi}{\hbar}\int|\langle v|\hat{H}|c\rangle|^2 \frac{2}{8\pi^3}\delta\,(E_c(\vec{k})-E_v(\vec{k})-\hbar\omega\,)\mathrm{d}^3k \qquad (4.13)$$

where $\langle v|\hat{H}|c\rangle$ is the transition matrix from states in the valence band (VB) to states in the CB and the integration is over the whole reciprocal space. For simplicity, the transition matrix $\langle v|\hat{H}|c\rangle$ can be considered as a "constant" (independent of k), and the absorption formula is then given by

$$\frac{2\pi}{\hbar}|\langle v|\hat{H}|c\rangle|^2 \cdot\int\frac{2}{8\pi^3}\delta\,(E_c(\vec{k})-E_v(\vec{k})-\hbar\omega\,)\mathrm{d}^3k \qquad (4.14)$$

where the second term is the joint density of states (JDOS) at energy $\hbar\omega$. Consequently, the optical absorption of a semiconductor is fundamentally determined by the transition matrix and JDOS. The JDOS is related to the density of states (DOS) of the CB and VB, contingent on the atomic orbitals that form the CB and VB.

Figure 4.4 shows the schematic band structures of the transitions responsible for the optical absorptions of three representative PV absorber materials, Si, GaAs, and lead halide perovskites. In terms of Si absorber, the

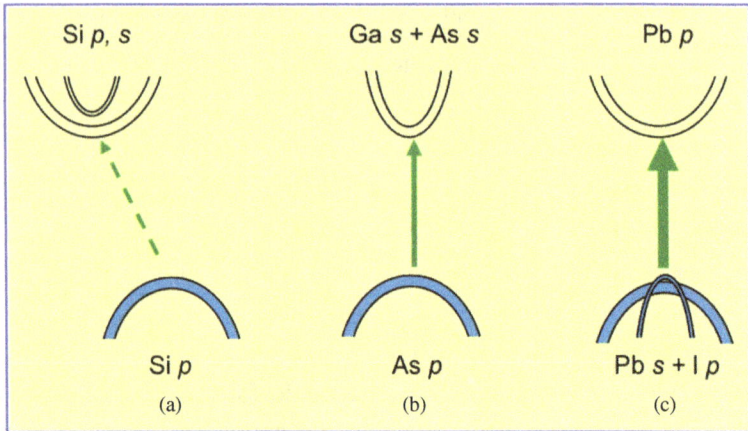

Figure 4.4.   The schematic optical absorption of (a) Si, (b) GaAs, and (c) MAPbI$_3$ perovskite. Reproduced from Ref. [6].

forbidden transitions of the indirect bandgap nature result in the low optical absorption. Both GaAs and lead halide perovskites (MAPbI$_3$ as an example, where MA = CH$_3$NH$_3$) have allowed transitions between band edges on account of their direct bandgap nature. The lower part of the CB of GaAs is derived from the dispersive Ga 4s states, whereas the one of MAPbI$_3$ is mainly composed of Pb 6p bands, because the s bands are more dispersive than the p bands. In addition, MAPbI$_3$ has a much higher JDOS, and therefore, stronger optical absorption than GaAs. It is seen that the optical absorption coefficient of MAPbI$_3$ is up to one order of magnitude higher than that of GaAs among the visible light range. As a result, the calculated maximum PCEs of MAPbI$_3$ perovskites are much higher than GaAs for any given thickness. This high optical absorption makes halide perovskites capable of achieving high PCEs with very thin absorber layers which take into consideration carrier concentration and minority carrier lifetime.

The carrier diffusion length is typically subordinate to the effective masses (m^*), non-radiative recombination, and scattering of carriers. The effective masses for electrons and holes are determined by the dispersion of the edges of CBs and VBs and can be approximately described by

$$m^* = \hbar^2 \left[\frac{\partial^2 \varepsilon (k)}{\partial^2 k} \right]^{-1} \tag{4.15}$$

where $\varepsilon(k)$ are the energy dispersion relation functions described by band structures. CB and VB edges with larger dispersions result in smaller effective masses. The CBMs of conventional GaAs solar cell absorbers (*p-s* semiconductors) are mostly contributed by cations and anion *s* orbitals, whereas the VBMs are mostly resulted from anions' *p* characters. High-energy-level *s* orbitals are more delocalized than low-energy-level p orbitals. Consequently, the electron effective mass is smaller than hole effective mass in the *p-s* semiconductors.

As discussed above, Pb^{2+} ions manifest $6s^2$ electron configurations in lead halide perovskites. The lower CB of Pb halide perovskites is mainly derived from the tenantless Pb *p* orbitals. The higher VB is mainly halogen *p* orbitals mixing with a small component of Pb *s* states. Therefore, the electronic structure of $MAPbI_3$ is significantly inverted contrasted with *p-s* semiconductors. Nevertheless, the *p* orbital of cation Pb has a much higher energy level than anion *p* orbitals. Thus, the lower CB of lead halides is expected to be more dispersive than the upper VB in conventional *p-s* semiconductors. Furthermore, on account of Pb, lone-pair *s* orbitals are close in energy to the filled I *p* orbitals, resulting in strong *s-p* coupling, making the upper VB of $MAPbI_3$ very dispersive. As a result, the effective masses are smaller for holes than for electrons in lead halide perovskites. Lead halide perovskites exhibit small electron effective masses though the conduction band edge is derived from Pb $6p$ orbital, owing to the strong SOC. The strong SOC in lead halide perovskites represents a unique advantage to achieve small carrier effective masses, leading to higher carrier mobilities, longer carrier diffusion lengths and less non-radiative recombination.

The defects as well as point defects, point defect complexes, surfaces, and GBs play critical roles in determining the device performance because they inevitably exist in solar cell absorbers. The formation of shallow-level point defects that provide free carriers is desirable, whereas the formation of deep level defects that cause non-radiative recombination is undesirable. The lead halide perovskites exhibit unique defect tolerance nature: first, the majority of the point defects formed in lead halide perovskites produce only shallow levels; second, the photogenerated electrons and holes may form large polarons, producing charge screening that generates a barrier for them from recombining at the deep-level defects. Fundamentally, the excellent defect tolerance of lead halide perovskites is attributed to the combination of high crystal symmetry, the inactive Pb $6p$ orbital, the Pb $6s$ lone-pair states, the ionic nature, and large atomic size. Among them, the inactive Pb $6p$ orbital and Pb $6s$ lone-pair states are the most important features.

Electronic dimensionality is a significant factor that determines the optoelectronic properties of semiconductors. The electronic dimensionality, defined by the connectivity of the electronic orbitals that comprise the CBM and the VBM, is better than the conventional structural dimensionality for explaining PV properties, such as bandgaps, carrier mobilities, defect levels, and device performances. It is proposed that a promising PV absorber should exhibit a high electronic dimensionality, as is the case for the current mainstream solar cell absorber materials (including Pb halide perovskites). It is noted that for many absorbers, their optoelectronic properties can be explained by their conventional structural dimensionality, owing to the strong correlation between their electronic and structural dimensionality. As the structural dimensionality reduces, the electronic dimensionality generally decreases, and the associated PV properties deteriorate. $CsPbI_3$, Cs_2PbI_4, Cs_3PbI_5, and Cs_4PbI_6 consist of the same elements Cs, Pb, and I, but have different structural dimensionalities, 3D, 2D, 1D, and 0D, respectively. The calculated bandgap (1.48, 1.90, 2.80, and 3.44 eV for the 3D, 2D, 1D, and 0D structures, respectively) increases with the decrease of the electronic and structural dimensionality. For 2D absorber materials, film growth with the orientation parallel to the layers are preferred for facilitating the effective carrier generation, transport, and collection. For the 1D and 0D structures, the bandgaps are too large and the band edges are too localized, making them not suitable for single-junction PV applications. It should be noted that a 3D structure is just a prerequisite for 3D electronic dimensionality. A 3D structure does not guarantee a 3D electronic dimensionality, because the latter is more advanced and describes the connectivity of only the orbitals that contribute to the band edges. The above statements suggest that the 3D structure is a prerequisite for high electronic dimensionality and for a promising absorber candidate.

4.1.4 *The Major Milestones of PSCs*

During the development from 3.8% to >25% PCEs, a number of milestones at the fronts of both device performance and basic mechanisms have been obtained. As for the case of infantile developments, due to the first and foremost requirement for achieving high cell efficiency, the fabrication of a high-quality perovskite film was a central issue. A great amount of effort went into the development of processes with a wide variation of processing methods and conditions to prepare high-quality perovskite films. Unfortunately, the high sensitivity of perovskites has made them difficult to find a general correlation between the perovskite film characteristics and performance. However, as more and more studies pertaining to the influences of perovskite film morphology on cell performance were undertaken, it became clear and comprehensible that a flat and dense (pinhole-free) perovskite absorber film with large grains and high crystallinity is a prerequisite for high efficiency. A variety of methods, such as anti-solvent dripping, solvent-assisted vapor annealing, inclusion of additives in precursor solution, and even the use of special annealing conditions, have been employed to achieve the required characteristics of perovskite films that culminate in high efficiency. Regardless, a high-quality perovskite film (pinhole-free, large grains) is the basic requirement for a high-performance device. Simultaneously, a lot of attention is also focused on the issues of anomalous J–V hysteresis and the important roles of different interfaces in device performance. Critical effects of the interfaces of perovskite with both electron and hole transport layers on device performance have been highlighted in many studies. Although there is no clear consensus on the origin of J–V hysteresis till now, a number of methods based on interfacial engineering have been established to either reduce or eliminate hysteresis. The use of organic electron transport layers and modification of the most widely used TiO_2 ETM layer surface with organic molecules have become popular interface engineering strategies to reduce J–V hysteresis. Although most recent studies suggest ion migration is responsible for such anomalous hysteresis, some other studies claim that traps at the interface and interfacial recombination cause the hysteresis.

In recent years, key developments have been reported in improving efficiency and stability simultaneously by compositional engineering of perovskites. Structural/intrinsic stability and long-term environmental stability of perovskites have become the most important issues at the present time. Current challenges touch on different recent and important developments, such as (i) compositional engineering of perovskites (mixtures of cations and anions that have resulted in concurrent enhancements in performance and stability), (ii) increasing potential of all-inorganic perovskites, (iii) exploring Pb-free perovskites, (iv) stability issues with perovskite and with other layers, and (v) challenges and potentials of scaling up perovskite PV.

4.1.5 *Processing Engineering of PSCs*

The performance of PSCs is determined by film qualities, i.e., crystallinity and morphology, which are dependent upon the film fabrication process. For example, highly efficient PSCs possess a dense microstructure without any pores. In this section, many proposed fabrication processes are introduced. In typical PSCs fabrication, the perovskite thin films are formed by low-temperature solution processes. The low formation enthalpy of MHP materials and the high ionic nature of their chemical bonding enable the utilization of cost-effective solution processes. Nevertheless, the crystallization processes of the perovskite films are completed within few seconds, limiting effective control over the crystallinity and orientation of the film. In solution processes, intermolecular interactions between the perovskite precursor and solvent molecules, as well as fabrication protocols, were found to have critical influences on the crystallization kinetics of the perovskite films.

The La Mer mechanism is applied to understand the crystallization processes of the perovskite films. Three different stages of the crystallization processes of perovskite films are demonstrated according to the La Mer mechanism. In stage I, the perovskite precursors exist as ions and molecules in the precursor solution until the concentration of the solution increases to a supersaturation level (C_s), which is due to the evaporation of the solvents in the spin coating and heat treatment of the film. As the concentration reaches C_s, in stage II, perovskite crystal nuclei start to form and grow with a supply of solute owing to a diffusion process. The nucleation continues until the concentration of the solution decreases below C_s, as the consumption of the solutes becomes faster than the evaporation of the solutes in stage III. In stage III, nucleation does not occur, and only the growth of the formed nuclei progresses until the precursor solution is depleted [7].

4.1.5.1 *One-step process*

In the one-step process, perovskite films are formed via the evaporation of the solvent from undried films that are deposited from a homogeneous solution containing all components. Various methods have been proposed to improve PV performance via the fabrication of uniform, well-crystallized, and void-free perovskite layers. The one-step process was first proposed by Kojima *et al.* (2009). The raw precursors, including PbI_2 and methylammonium iodide (MAI), are dissolved in a polar aprotic solvent such as DMF, DMSO, and GBL, following which the solution is directly cast onto the substrate using spin-coating. During the spinning process, the solvent evaporates; concurrently, the concentration of the dissolved precursor reaches the supersaturation state, thereby inducing the nucleation of perovskite or preperovskite materials. The as-dried films are then annealed at a low temperature in the range of 100–150 °C to control the nucleation and growth of the perovskite phase.

At an early stage, the conventional one-step process presented some critical issues that prevented the attainment of high efficiency, such as difficulty in fully covering the substrate surface with the perovskite layer and irregular grain growth due to the unsuccessful control of nucleation and growth. For example, in a well-used DMF solvent system, irregular rod or needle-like structures with poor coverage on the substrate were frequently observed. The high boiling point and low vapor pressure of DMF retard evaporation, leading to slow nucleation and rapid growth, thereby preventing the generation of a uniform grain structure. Therefore, research on the one-step process is focused on exploring the methods of quickly removing the solvent to control the rapid and uniform nucleation of perovskite mediates. Rapid solvent removal has been attained through various methods, such as solvent engineering, substrate heating, gas blowing, and vacuum treatment.

The solvent engineering method, also known as the antisolvent extraction method, was first reported by Cheng *et al.* The antisolvent should be miscible with the polar aprotic solvent but should not dissolve perovskite ingredients such as PbI_2 and MAI. They dripped chlorobenzene as an antisolvent to rapidly extract solvent from the cast film, thereby facilitating uniform nucleation. To date, many antisolvents, including toluene, hexane, ethyl acetate, and diethyl ether, have been explored. Many high-efficiency solar cells have been fabricated using this antisolvent extraction method. Moreover, antisolvent extraction can be applied to fabricate mixed halide compounds, such as $[CsPbI_3]_{0.05}[(FAPbI_3)_{0.85}(MAPbBr_3)_{0.15}]_{0.95}$. However, unlike $MAPbI_3$, the diffusivity of other ions such as Br is relatively high than that of I, which induces the formation of pinholes in the mixed halide perovskites. Furthermore, the molecular interaction between the perovskite precursor ions and solvent molecules in the solution is dependent on the composition of the added precursors. Therefore, a change in the composition of the perovskite results in variable precursor solubility and the solvent evaporation rate, inducing differences in the optimized antisolvent timing. The process window for the mixed halide perovskite was found to be fairly narrow for conventional one-step fabrication using a chlorobenzene antisolvent. Recently, anisole has been introduced as an antisolvent. Owing to a low evaporation rate and intermolecular interaction between the anisole and DMF/DMSO, anisole has an ultrawide processing window for the one-step fabrication of efficient PSCs, for which the small-cell efficiency (0.14 cm^2) is 19.76% (Figure 4.5) [8]. Further, the application of anisole enables the fabrication of uniform large-area perovskite films as large as 100 cm^2. The module efficiency of the anisole-derived PSC (100 cm^2) was 14%. Therefore, if a suitable antisolvent can be determined, antisolvent extraction can ensure a reproducible fabrication process for highly efficient PSCs as well as large-area devices.

In the antisolvent method, modification of the solvent and the addition of additives have been attempted to improve efficiency. For example, the use of a saturated mixture of methylamine in acetonitrile can improve the

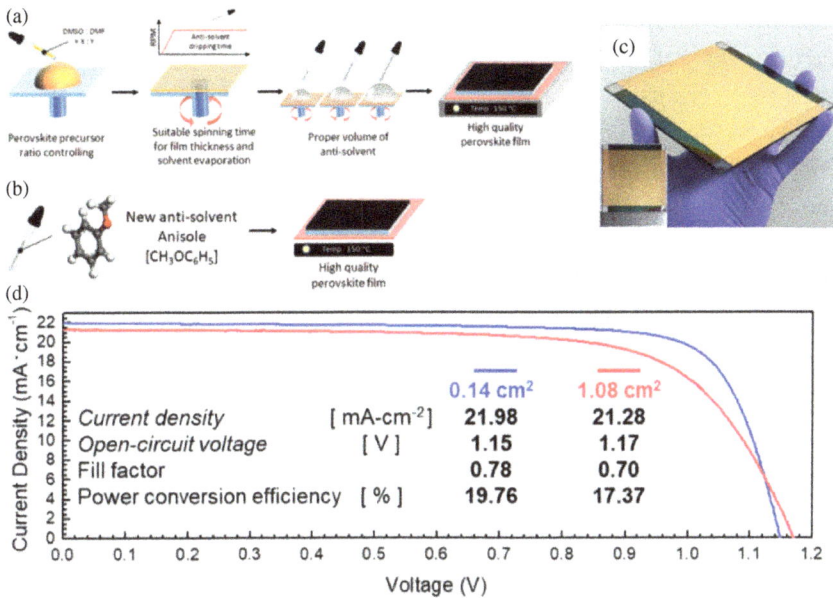

Figure 4.5. Schematic diagrams displaying procedure for one-step spin-coating of perovskite thin films using (a) conventional antisolvents and (b) anisole. (c) Photograph of the 10 cm × 10 cm perovskite solar module. Reproduced from Ref. [9]. (d) J–V curves of 0.14 and 1.08 cm² PSC devices fabricated by anisole as an antisolvent. Reproduced from Ref. [8].

perovskite microstructure. This modified solution allows the formation of a liquid "melt" of $MAPbI_3 \cdot xCH_3NH_2$, which results in the direct crystallization of the perovskite during spin coating process. The perovskite layer prepared by this method possessed fewer defects as well as a compact microstructure, yielding over 18% PCE. Most methods for achieving high efficiency over 20% are solvent engineering processes, which include a formation step for organic halide-Pb halide-DMSO intermediates and an antisolvent extraction step [10,11].

The simplest way to quickly evaporate solvents is to apply heat rapidly. Nie *et al.* developed the hot casting method by dripping a 70°C mixture of PbI_2 and MACl solution onto a 180°C heated substrate with spinning [12]. This process yielded exceptionally large millimeter-scale grains with reduced trap sites, leading to a high PCE of 18%. The heat energy enabled efficient evaporation of the solvent in the as-deposited film, which induced rapid nucleation. Moreover, the heat energy enhanced the diffusivity of ions, which is one of the reasons that this process yielded large-sized grains.

An N_2 gas air knife quickly blows out the solvent, leading to a high PCE of 20.26%. Because this gas-blowing method does not use toxic antisolvents such as toluene, chlorobenzene, and diethyl ether, this process is an environmentally friendly fabrication method. Recently, a cryogenic process was used to control the crystallization of perovskite films. This process enabled the decoupling of nucleation and crystallization phases by inhibiting chemical reactions in as-deposited films that were rapidly cooled by immersion in liquid nitrogen. The cooling was followed by conventional gas-blowing. This process initiated the formation of a uniform precursor seed layer, resulting in increased grain size, crystallinity, and low defect density. The best-performing cell attained a PCE of 21.4% [13].

Another effective method for evaporating the solvent is to use a vacuum. Li *et al.* devised vacuum flash-assisted solution processing (VASP) method, which could yield smooth perovskite films of high electronic quality over large areas [14]. The flash evaporation of a solvent produced a burst of perovskite precursor crystals and prevented dewetting, yielding homogeneous films without pinholes. The resultant PCE was 20.5%, owing to good crystallinity with reduced trap sites. Further, a combination of the gas blowing and vacuum process, termed the gas-flow-induced gas pump (GGPM) method, was developed. The maximum efficiency of a GGPM-manufactured cell was 20.44% [15].

4.1.5.2 *Two-step process*

The two-step process indicates a perovskite film fabrication procedure that yields a perovskite layer through the step-by-step deposition of each precursor layer, followed by a thermal interdiffusion process. The two-step process, or sequential deposition method, for the fabrication of PSCs involves (i) the deposition of Pb precursors, such as a Pb halide or Pb-based adduct, on a substrate, (ii) exposure to a liquid, vapor, or solid of an organic salt, and (iii) a diffusion-driven formation of perovskite

materials that is usually enabled by heat treatment. This process is benefi-
cial for the moderate control of reactions between the Pb halide and
organic salt, which induces the crystallization of perovskite materials.
Recently, a detailed path for the two-step process for MAPbI$_3$ formation
has been proposed by Ummadisingu *et al.* [16]. As shown in Figure 4.6,
the formation behavior adopts a sequential path involving (i) the nucleation and growth of PbI$_2$, (ii) the intercalation of MAI and structural reorganization to form MAPbI$_3$, (iii) Ostwald ripening, in which the perovskite
from the mesoporous TiO$_2$ is transported to the capping layer, and (iv)
further Ostwald ripening during dipping times in which the perovskite is
transformed from small crystals in the capping layer to large crystals.
Severe Ostwald ripening is detrimental to a high PCE. Each step should
be carefully controlled to obtain highly crystalline and dense morphology
to attain high efficiency.

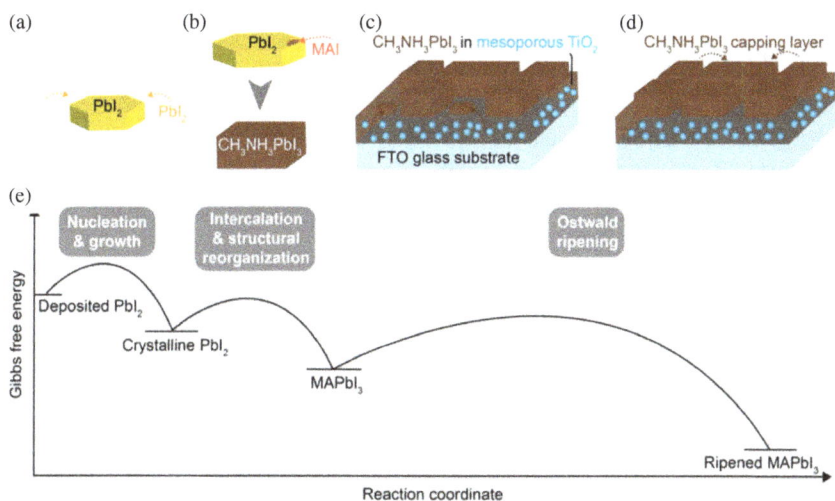

Figure 4.6. Schematic depiction of the stages of the reaction in sequential deposition.
Dashed arrows indicate mass transfer. (a) Nucleation and growth of PbI$_2$. (b) Intercalation
of MAI and structural reorganization to form CH$_3$NH$_3$PbI$_3$ perovskite. (c) Ostwald ripening, in which the perovskite from the mesoporous layer is transported to the capping layer.
(d) Further Ostwald ripening at longer dipping times in which the perovskite from the
small crystals in the capping layer is transported to larger crystals. (e) Gibbs free energy
as a function of the reaction coordinate. Reproduced from Ref. [16].

The two-step method was first proposed by Liang *et al.* to fabricate MAPbI$_3$ films through the deposition of PbI$_2$ on a glass or quartz substrate and subsequent dipping into an MAI solution [17]. For PSCs, Burschka *et al.* slightly modified this method for the deposition of PbI$_2$ as well as the formation reaction for increasing the loading of the perovskite absorber on the mesoporous TiO$_2$ film [18]. The spinning time for PbI$_2$ casting was as short as 5 s, and subsequently, the PbI$_2$ films were prewetted by dipping them in 2-propanol for 1–2 s before dipping in a solution of MAI and 2-propanol. This process led to a PCE of 15.0% with good reproducibility. The PCE was further increased to 17.01% by adopting the size-controlled cuboidal MAPbI$_3$ proposed by Im *et al.* The cuboid MAPbI$_3$ was prepared using the two-step method, and its size was controlled by changing the concentration of MAI and the exposure time of PbI$_2$ to the MAI solution before spin-coating. The average cuboid size decreased with the concentration of MAI.

Mixed halide materials, such as a mixture of FA and MA cation-based perovskite materials, were successfully synthesized using the two-step process, which extended the optical absorption onset further into the red region for the enhancement of light harvesting. The mixed cation-based MA$_{0.6}$FA$_{0.4}$PbI$_3$ PSC yielded a PCE of 14.9% by virtue of superior light harvesting and charge-collection efficiency [19]. A highly efficient PSC was realized by exploiting the intra-molecular exchange process (IEP), wherein the PbI$_2$·DMSO film was deposited as a first step, followed by the deposition of FAI and post annealing [20]. During annealing, DMSO was exchanged with FAI, consequently forming FAPbI$_3$ with large-grained dense microstructures and flat surfaces without residual PbI$_2$. In addition, mixed FAPbI$_3$/MAPbBr$_3$ was synthesized by IEP. The resultant cell efficiency was 20.1%, certified by the standardized method in a PV calibration laboratory.

4.1.5.3 *Adduct intermediated process*

Although extensive efforts have been invested to improve the efficiency of PSCs, the standard deviation of the average PCE is fairly large, which indicates that the reproducibility of high-efficiency PSCs is a persisting challenge. The presence of the intermediate phase was initially observed by Jeon *et al.* [18], which was later found to be a Lewis acid–base adduct composed of perovskite precursors and Lewis base solvent molecules [21]. The Lewis base adduct of MAI·PbI_2·DMSO was formed by spin-casting an equimolar MAI, PbI_2, and DMSO solution dissolved in DMF solvent. The diethyl ether quickly extracted the DMF solvent, which formed the adduct of MAI·PbI_2·DMSO. The adduct film converted into $MAPbI_3$ by simple post-annealing. The resultant perovskite film showed more uniform morphology compared with the perovskite film fabricated using conventional PbI_2. Chiefly, the adduct method yielded a high reproducibility, i.e., a narrow standard deviation with average PCE of 18.3% from 41 cells and the best PCE of 19.7%. DFT calculations showed a higher interaction energy of the NMP with the FA cation, supporting the experimental results on the formation of a strong coordinative bond with FAI and NMP. Owing to the excellent uniformity and reproducibility of $FAPbI_3$, the resulting PSC attained a high PCE of 20.19%. Moreover, in this report, criteria for selecting Lewis bases were suggested: (1) a high hydrogen bond-accepting ability and low hydrogen bond-donating ability, (2) a sterically accessible electron-donating atom, and (3) matching the hardness and softness of the Lewis acid and base.

Recently, a new fabrication method for perovskite layers based on an adduct compound has been reported. Simple face-to-face contact without pressure, termed the bifacial stamping process, enabled the low-temperature phase transition of $FAPbI_3$ and the effective interface modification of $MAPbI_3$, yielding superior PV performances for both perovskites [22]. This result was explained in terms of the content of the DMSO reservoir in the $MAPbI_3$ adduct film, which induced the phase transition of $FAPbI_3$ and high-quality $MAPbI_3$. During the bifacial stamping, a DMSO-mediated ion exchange reaction took place, which facilitated ion transport from the Lewis base reservoir in the $MAPbI_3$ adduct film, consequently resulting in the fast phase transition of $FAPbI_3$ at a low temperature. The adduct-intermediate process is promising for preparing dense perovskite films with large grains, which is one of the requirements for achieving high efficiency.

4.1.6 *Over Shockley–Queisser Limit (Tandem solar cells)*

The formation of a tandem configuration, in which two or more photosystems with different bandgap energies are combined in series or parallel, is the only feasible approach to achieve conversion efficiencies above the Shockley–Queisser limit. This has been proven by the high efficiencies observed in multijunction solar cells. In a double-junction solar cell (or tandem solar cell with two photosystems), for example, the top cell with a higher bandgap energy absorbs high-energy photons, whereas the bottom cell with a lower bandgap energy absorbs low-energy photons which have transmitted through the top cell [23]. In the tandem configuration, a wide range of solar radiation can be effectively shared by the two photosystems while minimizing the thermalization loss of the photons' excess energy. Therefore, the much higher conversion efficiencies for tandem solar cells can be expected when compared with their single-junction counterparts.

In terms of the device structure, tandem solar cells can be classified as monolithic two-terminal, mechanical four-terminal, and optical four-terminal devices. In a two-terminal monolithic tandem solar cell, two subcells are directly connected in series with a tunnel junction or recombination layer, in which the overall tandem current is limited by the subcell exhibiting the lowest photocurrent. Therefore, it is very important that each subcell generates the same photocurrent by efficiently sharing the incident photons in order to obtain a high matching photocurrent of the tandem solar cell. The photocurrent in the subcells can be successfully matched by controlling the optical properties, such as the bandgap/transmittance of the layers, light scattering, and up/down-conversion of incident light. Most perovskite-based tandem solar cells including the champion device are prepared using a crystalline Si bottom cell. The optimum bandgap energy of the perovskite top cell that perfectly matches that of Si (1.12 eV) is 1.68 eV, which is somewhat higher than that for an optimal PSC and requires a mixed halide system incorporating Br. To improve the photovoltage of tandem solar cells, the photovoltage of each subcell has to be improved independently. Therefore, numerous approaches that have been applied toward improving the photovoltage of single-junction PSCs can be applied to large-bandgap perovskites, such as the passivation of defects using additives. However, the relatively large V_{oc}-deficit (i.e., E_g–V_{oc}) and low photostability of mixed halides (e.g., phase segregation) have yet to be addressed [24,25]. The *FF* values of tandem solar cells can be increased by either reducing the dark current and/or shunt conductance or by reducing the series resistance of the multiple layers and the interfacial charge transfer resistance between them [26]. In addition, developing more reliable characterization methods including EQE/*J–V* measurements and electron dynamics for tandem solar cells and their subcells is an important research objective.

The first successful demonstration of a perovskite/Si tandem solar cell with a certified conversion efficiency of 23.6% was provided by Bush *et al.* A perovskite top cell with an E_g of 1.63 eV ($Cs_{0.17}FA_{0.83}Pb(Br_{0.17}I_{0.83})_3$) was directly deposited onto a polished Si heterojunction bottom cell with a textured backside [27]. A number of additional interfacial layers were deposited to fabricate the functioning tandem solar cell, such as the SnO_2/ZTO layer used to protect the perovskite cell from the plasma damage that occurs during the deposition of the ITO window layer and the silicon nanoparticle layer on the textured backside used to improve the near IR photoresponse of the Si bottom cell. More noteworthy than the high conversion efficiency is that the tandem solar cell passed a damp heat test designed for crystalline silicon modules (IEC 61215), suggesting that the stability of perovskite-based tandem solar cells can be significantly improved with the aid of an appropriate encapsulation agent, such as an inorganic TCO layer, EVA film, and the glass cover sheet. One can expect significantly higher photo responses with textures on both sides of the Si bottom cell, especially toward the near-IR region; however, directly fabricating a perovskite top cell on top of a textured Si bottom cell is quite challenging. Sahli *et al.* reported a vacuum-solution hybrid approach used to fabricate a monolithic perovskite/Si tandem solar cell with a fully textured Si bottom cell [28]. Because of the effective light scattering of the textured top surface, a very high overall tandem current density close to 20 mA/cm² and high conversion efficiency of >25% could be achieved. As a result of the continued research efforts of several research groups, the highest efficiency of perovskite/Si tandem solar cells has recently reached

29.1%. More recently, several notable papers on highly efficient and stable perovskite Si tandem solar cells have been published. Hou *et al.* demonstrated that solution processes can be used to prepare perovskite top cells on textured Si bottom cells and reported a certified efficiency of 25.7% with good device stability. Xu *et al.* reported that the phase segregation observed in perovskite top cells using mixed-halide (i.e., Br/I) compositions can be successfully suppressed by the incorporation of Cl⁻ ions and, thus, by the formation of triple-halide compositions.

Given the rapidly increasing conversion efficiencies of perovskite-based tandem solar cells, it is reasonable to expect ultrahigh efficiencies of >30% in the near future. Then, what are the key scientific issues or the technical challenges? Bandgap engineering is probably one of the most important issues, which not only involves tuning the bandgap to the desired value but also optimizing the performance of the single-junction device with a specific bandgap energy. The typical bandgap energies of perovskite required for tandem solar cells are 1.2–1.25 eV for the bottom cell of perovskite/perovskite tandem devices, 1.4–1.5 eV for the middle cell of triple junction devices, 1.65–1.7 eV for the top cell of perovskite/Si(CIGS) tandem devices, 1.75–1.8 eV for the top cell of perovskite–perovskite tandem devices, and 1.9–2.0 eV for the top cell of triple junction devices.

4.2 Solid-State Light-Emitting Devices

Solid-state light-emitting devices such as LEDs and lasers play important roles in modern industrial application and daily life. They have the potential to address urgent challenges not only relating to general lighting and display applications but also relating to energy saving and greenhouse gas emissions [29]. Through development of materials science, a variety of light-emitting materials have been demonstrated since the first demonstration of visible-spectrum LEDs based on gallium arsenide phosphide (GaAsP) over five decades ago. This has led to the evolution of various LED technologies: inorganic semiconductor LEDs, organic LEDs (OLEDs), polymer LEDs (PLEDs), quantum-dot LEDs (QLEDs), and recently, perovskite LEDs (PeLEDs), as described in detail in what follows.

4.2.1 Perovskite light-emitting diodes

With the development of solid-state lighting technology, efficient light sources that combine high brightness, wide range, and good stability are in high demand for next-generation lighting and displays. MHPs are emerging as promising luminescent materials due to their versatility for desirable light emission manipulations. This is because the optical activity of the metal halide material depends on the metal halide structural unit and the organic ions or coordinated organic ligands. The different assembly of metal halide units and organic parts can enable versatile light emissions, in which the most typical members are lead halide perovskites. Impressively, the external quantum efficiency of the LHP based LEDs has improved significantly from 0.1% to over 20% in just 5 years. Nonetheless, the structural lability and toxicity of the LHPs are still the critical drawbacks that need to be addressed for practical applications.

4.2.1.1 Comparison of PeLED with QLED and OLEDs

The comparison of traditional II–VI quantum dots (QDs) and organic materials with perovskites in light-emitting applications is listed (Table 4.2). QDs tunable bandgap, as governed by the quantum size effect, also poses challenges, for it mandates high monodispersity in the population of QDs.

Table 4.2. Comparison of PeLEDs, QD LEDs, and OLEDs.

	PeLEDs	QD LEDs	OLEDs
FWHM (nm)	~20	~28	~70
$ELQY^{red}$ (%)	13	20.5	29
L_{max}^{red} (cd/m^2)	82 W sr^{-1} m^{-2}	42,000	1,000,000
$ELQY^{green}$ (%)	14	14.5	27.5
L_{max}^{green} (cd/m^2)	9,120	10,000	73,100
$ELQY^{blue}$ (%)	1.5	10.7	36.7
L_{max}^{blue} (cd/m^2)	2,480	4,000	10,000
Operational stability (h)	10	100,000	600,000
Color tunability	400 nm-near IR	400 nm-near IR	400 nm–700 nm

A precisely tailored spectrum can be used to generate a specific color temperature of white light. The quality of white light is measured via the correlated color temperature (CCT) and color rendering index (CRI). For lower-CCT illumination at high temperature, high luminous efficiency and color quality are hard to achieve: the red luminophores require a very narrow emission line width. Conventional red phosphors have emission spectra that are very broad (>60 nm for the full width at half-maximum (FWHM)). The narrow spectral emissions (~30 nm FWHM) of QDs offer more selective optical down-conversion of a portion of a backlight's blue emission into red-shifted light, leading to a CRI of >90% and a superior CCT of 2,700 K while maintaining a high luminance efficiency [30]. This allows QDs to increase color quality while lowering power consumption in solid-state light sources. QDs can also be utilized as backlights in high-color-quality liquid-crystal displays.

Nonetheless, QDs also still face some challenges. The importance of monodispersity in QDs for electronics arises because a rough energy landscape impedes transport and enhances recombination due to the presence of small-bandgap inclusions. Charging can occur when under current injection, which retards further injection and thus reduces EL efficiency. The time scales associated with QD charging range from minutes to days [31]. Although the PLQY of QD solution routinely often exceeds 95%, it is often much lower for QDs in solid films (often by 1 order of magnitude). Embedding QDs into an insulating polymer matrix decreases QD emission quenching; however, the electrical conductivity of these QD-polymer composites is impeded by the low conductance of the wide bandgap polymers. Förster resonant energy transfer (FRET) between closed-packed QDs can also reduce the PLQY of QD films.

While QDs begin to be deployed in applications such as backlights in displays, OLEDs are already a multibillion-dollar industry with applications in displays, lighting, and consumer devices. Most applications of OLEDs could benefit from a solution-processed approach (instead of the prevailing vacuum sublimation process), where roll-to-roll solution-processing and inkjet printing could potentially be employed [32]. To achieve this, new high-performance active materials that are solution-processable would accelerate adoption. Efficiency roll-off represents another remaining area for improvement in OLEDs. This corresponds to both a lowered PCE as well as added electrical stress to achieve a given brightness, which works against device lifetime.

Perovskites offer pathways to color-saturated emissions in view of their narrow emission line widths (<30 nm) and manifest the possibility for future high-color-purity emitters. Ready wavelength tuning and efficient charge injection/transport property with perovskites have been studied as promising features to enable next-generation light emitters. PeLEDs also have shown sub-bandgap turn-on voltages, implying the possibility of achieving low operating voltages and high-power conversion efficiencies in display applications. PeLEDs that can be processed from the solution phase have the potential to be inkjet printed. The processing temperatures may be quite low (less than 150 °C), making fabrication compatible with flexible and light-weight plastic substrates.

4.2.1.2 *The device architecture of PeLED*

The demonstration of electroluminescence in halide perovskites was first observed in layered perovskites at cryogenic temperatures in the early 1990s and later at room-temperature (RT) by incorporating a specially modified quaterthiophene dye in the hybrid perovskite sheet [33]. The device architecture of PeLED can be classified into multi-layered and single-layered device structures. A typical multi-layered PeLED device consists of a front transparent electrode (typically FTO or ITO), an *n*-type hole-blocking layer (HBL), a *p*-type electron-blocking layer (EBL), a perovskite emitter, and a back electrode. The perovskite active layer is sandwiched between the HBL and the EBL to form a double-heterojunction structure in order to confine the injected charges for better light emission. Under applied voltage, the electrodes inject charge carriers through the transporting layers into the perovskite active layer, where they recombine radiatively and emit light. In general, multi-layered PeLEDs have two main device geometries: the conventional (HBL/perovskite/EBL) and inverted (EBL/perovskite/HBL) configurations. These configurations are adapted from the basic device structures employed in solar cells. The typical schematic diagrams of the device geometries (conventional and inverted) show some of the reported carrier blocking layers. Alternatively, the single-layered PeLED device is composed of a perovskite composite sandwiched between an anode and a cathode. The most commonly used perovskite emitters are based on $MAPbX_3$ and $CsPbX_3$ (X = Cl, Br, I or a mixture of them) family of halide perovskites, although recent reports also demonstrated the use of mixed-cation PeLEDs [34].

4.2.1.3 *Perovskite for LEDs*

Modulating dimensionality perovskite: In the early 1990s, the Nurmikko and Saito groups reported LEDs based on layered perovskites. The authors used a layered perovskite composition $(C_6H_5C_2H_4NH_3)_2PbI_4$ and measured the electroluminescence (EL) spectrum at liquid nitrogen temperatures of ~77 K. The increased scattering rate due to phonons at higher temperatures explained the observed thermal quenching.

In 2014, Friend *et al.* reported room-temperature PeLEDs based on $CH_3NH_3PbI_{3-x}Cl_x$ with an EQE of 0.76% and radiance of 13.2 W sr^{-1} m^{-2} at a current density of 363 mA cm^{-2}. Lee and co-workers reported PeLEDs based on $MAPbBr_3$ with a current efficiency (CE) of 0.577 cd A^{-1}, an EQE of 0.125%, and a maximum luminance of 417 cd m^{-2}. Multi-colored LEDs were demonstrated by leveraging the mixture of halides. These early studies leveraged synthetic control over the perovskite structure and composition, with work by Huang, Wang, and colleagues demonstrating tunable electroluminescence across a wide range of the visible spectrum from ca. 500 to 800 nm using a perovskite with the nominal composition $(C_{10}H_7CH_2NH_3)_2(CH(NH_2)_2)Pb_2Br_{7-x}I_x$ [35]. The perovskite thin films were thin enough to confine electrons and holes for bimolecular recombination, enhancing electron–hole capture and increasing radiative recombination. However, making pinhole-free films containing crystalline particles and having overall thickness less than 50 nm was still a challenge. One method to overcome this film-formation challenge is to blend the perovskite precursors with a polymer, that is also soluble in the solvents needed to process the metal halide and ammonium halide salts. For example, Li *et al.* blended perovskites with insulating polyimide and deposited uniform thin films. The perovskite grains provided for charge transport and light emission, while the dielectric polymer filled the surrounding pinholes and thus filled potential shunt paths. A similar approach with mixed perovskite/poly(ethylene oxide) films also facilitated perovskite film formation, producing LEDs with relatively high brightness (4,064 cd m^{-2}). As the presence of metallic Pb atoms in MAPbBr$_3$ has been offered as a cause of trap-induced non-radiative recombination, Lee and co-workers addressed the film non-uniformity issue by providing excess MABr, reducing grain size and suppressing exciton quenching. The spatially confined MAPbBr$_3$ grains (average diameter ≤100 nm) were formed by an NC pinning process with additional TPBi (2,2′,2″-(1,3,5-benzinetriyl)-tris(1-phenyl-1H-benzimidazole)) dissolved in the solvent used for the antisolvent quench. PeLEDs based on this approach had a maximum current efficiency of 43 cd A^{-1} corresponding to an EQE of 8.5%. Park *et al.* further investigated the introduction of TPBi diluted in a volatile non-polar solvent during the perovskite NC pinning process.

Further exciton and carrier confinement was achieved using low-dimensional perovskites, as reported by Yuan *et al.*, who used a perovskite composed of a series of differently quantum-size-tuned grains that funneled photoexcitation to the lowest-bandgap emitters in the mixture [36]. Quasi-2D perovskites with a composition of PEA$_2$(CH$_3$NH$_3$)$_{n-1}$Pb$_n$I$_{3n+1}$ (PEA = C$_8$H$_9$NH$_3$) were used to tune the average layer numbers $\langle n \rangle$ by changing the ratio between PEAI and CH$_3$NH$_3$I. Transient absorption (TA) and time-resolved PL spectroscopy enabled characterization of transport and recombination dynamics at ultrafast time scales. These studies of excitation dynamics revealed how multi-phased perovskite materials channel energy across an inhomogeneous energy landscape, concentrating excitons into the smaller-bandgap emitters (larger n). PeLEDs with $\langle n \rangle = 5$ perovskite exhibited the best performance within this study, with an EQE of 8.8%, radiance of 80 W sr^{-1} m^{-2}, and an emission wavelength of 750 nm.

Colloidally-synthesized perovskite nanocrystals: Kovalenko *et al.* pioneered all-inorganic cesium lead halide (CsPbX$_3$; X = Cl, Br, and I) NCs, which exhibited excellent optical properties with tunable bandgap and high PLQY [37].

Recently, PeLED devices have been fabricated with an ITO/PEDOT:PSS/halide perovskite/B$_3$PYMPM/LiF/Al configuration as shown in Figure 4.8 [38]. Here, PEDOT:PSS acts as the HTL, B$_3$PYMPM is the ETL, LiF is the electron-injection layer, and ITO and Al are the anode and

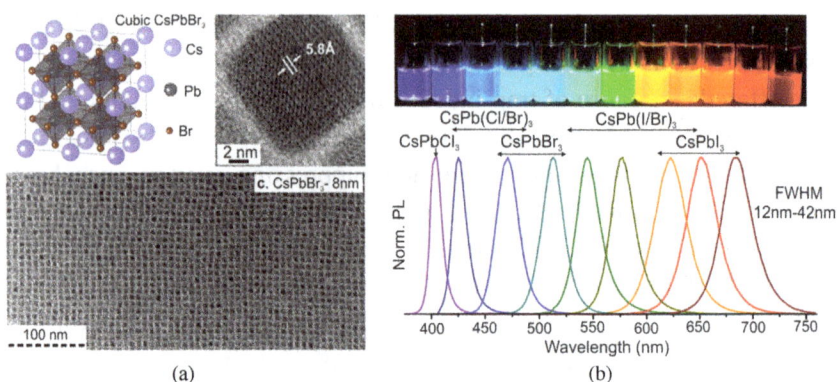

(a)   (b)

Figure 4.7. (a) Schematic representation of the cubic perovskite lattice, typical transmission electron microscopy (TEM) images of $CsPbBr_3$ NCs. (b) Colloidal solutions in toluene under UV lamp (λ = 365 nm), representative PL spectra (λ_{exc} = 400 nm for all but 350 nm for $CsPbCl_3$ samples. Reproduced from Ref. [37].

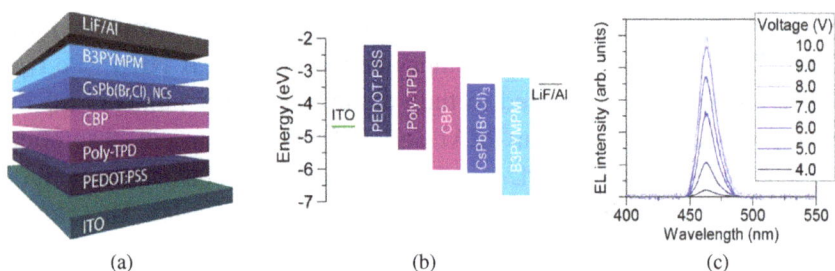

(a)   (b)   (c)

Figure 4.8. EL characteristics of a blue PeLED. (a) Device structure of the PeLED, (b) EL spectra of the PeLED at various applied voltages, and (c) EL spectra recorded at 5 V and different times. Reproduced from Ref. [38].

cathode, respectively. A record-breaking EQE (20.3%) can be achieved with such a configuration by incorporating a thin layer of insulating PMMA between the perovskite layer and ETL [39]. A PeLED device with record-breaking 20.3% EQE and 80% PLQY is fabricated using a compositionally graded quasi-core/shell $CsPbBr_3$/MABr perovskite layer. Here, MABr passivates the non-radiative defect sites, increases the PL life-time and the surface coverage, and balances charge-injection. In organic–inorganic hybrid perovskites, an interesting strategy to defect passivation is to use molecules with high affinity for defect sites. For example, a high EQE of 21.6% is achieved for a $FAPbI_3$-based NIR PeLED by using 2,20-[oxybis(ethylenoxy)]diethylamine to passivate defects.

Besides the passivation of defects, another successful approach to enhance EQE is to use low-dimensional perovskite structures such as nanocrystals, platelets, quasi-2D/3D nanostructures, and QDs. The EQE of PeLEDs constructed using bulk perovskite layers or films is limited by the thermal dissociation of excitons, which suppresses monomolecular or geminate-type excitonic recombination. Further, the bimolecular recombination kinetics in halide perovskites follow a quadratic relation with the charge carrier density, i.e., n_2. Therefore, the radiative recombination in such perovskite layers becomes dominant only at higher charge densities. Nanostructure perovskites lift these problems by enhancing the monomolecular radiative recombination through carrier confinement. For example, a red-emitting PeLED with a high EQE (21.3%) is constructed by using anion-exchanged perovskite QDs (Figure 4.9) [40]. Despite these approaches to improve the performance of PeLEDs, these devices are still behind commercial LEDs due to the poor stability of perovskites in diverse environments.

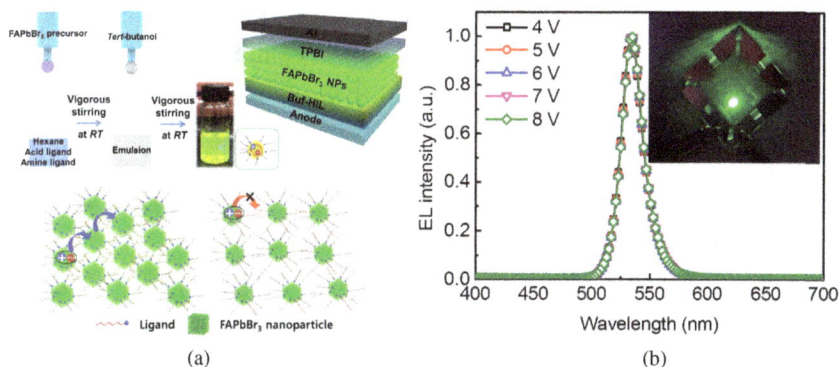

Figure 4.9. Schematic representation of the synthesis of $FAPbBr_3$ NPs using emulsion system at room temperature, (a) device architecture of $FAPbBr_3$ NP-LEDs, (b) schematic illustration of luminescence efficiency and charge injection/transport capabilities in $FAPbBr_3$ NP films with short ligand (left) and long ligand (right). (b) EL spectra at different applied bias and photograph (inset) of $FAPbBr_3$ NP LEDs using butylamine. Reproduced from Ref. [41].

Organic–inorganic hybrid-based NCs have also been advanced in LEDs. The performance of formamidinium lead bromide perovskite (FAPbBr$_3$) nanoparticles was improved with the aid of surface engineering. Surface ligand in NCs have improved the charge injection in LEDs, achieving a current efficiency of 9 cd/A, an important milestone (Figure 4.9) [23,41]. This has led to PLQYs exceeding 60% without using complex post-treatments or multilayers, with a current efficiency of 15.5 cd/A. Self-assembly of 3D FAPbBr$_3$ NCs with graded size (mixed microplatelets of octylammonium lead bromide perovskites) were synthesized by Chin *et al.*, enabling an energy cascade that yields high efficiencies in green-emitting LEDs. Transient optical spectroscopy revealed an energy cascade from high-bandgap 2D (OA)$_2$(FA)$_{n-1}$PbnBr$_{3n+1}$ platelets to progressively lower-bandgap FAPbBr$_3$ NCs, giving rise to high luminescence efficiency. Mesoscopic thin films comprising large plate-like domains of (OA)$_2$(FA)$_{n-1}$PbnBr$_{3n+1}$ were sandwiched between electron and hole transporting layers that enabled an EQE of 13.4% and 56,000 cd m^{-2} luminance.

4.2.1.4 *Stability of PeLEDs*

The device stability in QD LEDs is about 100–1,000 h when operated at relatively modest brightness (1,000 cd m^{-2}). In contrast, the device stability of state-of-the-art OLEDs is in the range of 10,000 to 100,000 h. Instability in early generation of OLED was also triggered by dark-spot effects; however, device encapsulation and optimization enabled the community to overcome these issues. The stability of lead halide-based PeLEDs remains a major issue in this early-stage materials platform. Here we discuss this challenge, along with recent studies toward increasing device stability.

Status and mechanisms underpinning stability in PeLEDs: PSCs' device lifetimes now exceed 1,000 h under continuous operation while biased at the maximum power point. In contrast, PeLEDs are at a less mature stage from a reliability point of view as devices often have subhour consistent operation under continuous electrical potential. Before degradation becomes observable at a brightness level higher than 100 cd m^{-2}. Some issues of PeLEDs are similar to that of PSCs. From PV studies, it became clear that some compositions of thin films lack chemical and structural stability, especially in the presence of moisture or heat. The strong ionic character of perovskites is linked to phase segregation and ion migration as well as to various electrochemical reactions (Figure 4.10).

In early 2014, stability issues of perovskites for PV gained substantial attention. Fundamental studies on instability in perovskite thin films in the presence of moisture and the surrounding environment were reported by Niu *et. al* [42]. Device fabrication needs to be carried out under controlled atmospheric conditions and with a humidity of <1% as shown by Gratzel and co-workers [43]. The degradation process was proposed to follow this route:

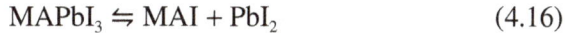

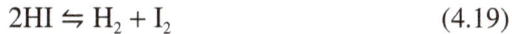

$$MAPbI_3 \leftrightharpoons MAI + PbI_2 \tag{4.16}$$

$$MAI \leftrightharpoons MA + HI \tag{4.17}$$

$$4HI + O_2 \leftrightharpoons 2I_2 + 2H_2O \tag{4.18}$$

$$2HI \leftrightharpoons H_2 + I_2 \tag{4.19}$$

In reaction (4.16), $MAPbI_3$ decomposes into MAI and PbI_2 in the presence of H_2O. In this process, the equilibrium constant is $k(1) = c(MAI)$ and

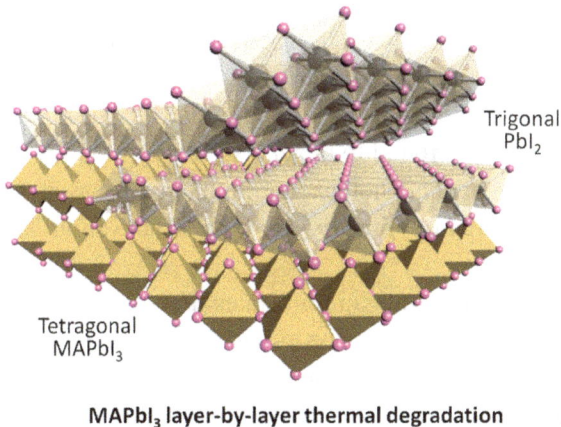

MAPbI₃ layer-by-layer thermal degradation

Figure 4.10.   Structural degradation in $MAPbI_3$. Reproduced from Ref. [23].

the Gibbs free energy is $\Delta G(1) = -RT \ln k(1)$. In reaction (4.17), MAI solution decomposes into MA solution and HI solution. Then the equilibrium constant is $k(2) = c(MA) \times c(HI)/c(MAI)$ and Gibbs free energy is $\Delta G(2) = -RT \ln k(1)$. In reaction (4.17), there are two ways for HI to react. Reaction (4.18) represents the redox reaction and can move forward readily. Reaction (4.19) represents the photochemical reaction, and the experimental evidence revealed that HI can easily decompose into H_2 and I_2 under optical excitation. *Ab initio* molecular dynamics simulations further predicted surface reconstruction and light-assisted formation of hydrated species of $MAPbI_3 \cdot H_2O$, a finding that agrees with the observation that exposure to water and light accelerates the degradation of perovskites [44].

Charge trapping at the interface between perovskites and charge extraction materials are responsible for irreversible degradation due to moisture [45]. To understand charge-driven degradation mechanisms, it was important to develop controlled stability experiments both for commonly used $MAPbI_3$, which is known to form structurally distorted tetragonal crystals, and for mixed perovskite materials having more enhanced structural stability. It was found that the mixed perovskite $MA_{0.6}FA_{0.4}PbI_{2.9}Br_{0.1}$ still degraded, though its degradation speed was slower than that of conventional $MAPbI_3$. It was further shown that the irreversible degradation of the perovskites was triggered by trapped charges.

Oxygen, in addition to moisture, can also trigger degradation [46]. Darkening in PL also involved interaction with photogenerated carriers. The dynamics of PL in $MAPbI_3$ films were studied under continuous photoexcitation. The activation and darkening of the PL intensity was observed, and this was found to depend strongly on optical pump characteristics. When the optical photoexcitation intensity was increased, the PL of perovskites increased linearly, and both activation and darkening dynamics increased.

Photoactivation fills subgap trap states that otherwise limit the PL of the perovskite. These states can be either shallow or deep, and photoactivation is more efficient in the presence of O_2 than it is with N_2. When and if the photoactivation process is actually related to the filling of charge traps, oxygen plays a role in the process. Perovskite films under prolonged optical excitation may photodarken, a fact attributed to the decomposition of the perovskites under excitation. The hydrated compounds are easily formed upon photoexcitation, and this leads to a weakening of hydrogen bonds between PbI_6 and organic cations. This ultimately promotes the formation of complexes between H_2O and PbI_6. Further research is required to understand the mechanisms of H_2O and O_2 in the photoactivated processes.

The mechanistic process of degradation has been studied using super-resolution luminescence microspectroscopy [47]. The degradation of the $MAPbI_3$ crystal structure starts locally due to the collapse of the layered PbI_2 structure; and then spreads across the crystal. The migration of MA ions can distort the perovskite lattice, and this leads to the loss of the crystal structure.

In addition to instabilities in perovskite materials, device instabilities can also come from interfacial layers and electrical stress during operation of device. Ionic processes, such as halide segregation, can destroy perovskites and generate defects, some of which are voltage-induced while others are spontaneous. Some examples include the corrosion/oxidation of electrodes, degradation of charge-transport layers, and the formation of charge-accumulated interfaces and p–i–n junctions. Furthermore, the diffusion of metallic species from electrodes, the origin of which is still unclear, can cause a significant decrease in device efficiency and accelerate the degradation of PeLEDs.

Strategies to improve LED stability: To improve the stability of PeLED, several approaches have been used. Inorganic $CsPbX_3$ perovskites have been shown to improve LED stability compared with MA-based perovskites, a finding associated with their enhanced thermal, structural, and chemical stability. The high thermal stability of $CsPbBr_3$ may improve the operational stability of PeLEDs by suppressing degradation caused by thermal stresses, including Joule heating. Schaller and co-workers reported temperature-dependent PL below 450 K, which was strongly affected by the halide composition [48].

Another strategy to improve perovskite stability involves the use of layered perovskite structures. $(PEA)_2(MA)_2[Pb_3I_{10}]$ has shown increased stability against moisture, attributed in part to the hydrophobic nature of the PEA layers. Using a density functional theory (DFT) approach, Quan *et al.* proposed that the introduction of PEA layers can increase quantitatively van der Waals interactions. These provide an increased formation energy corresponding to improved materials stability. Recently, quasi-2D perovskite/poly(ethylene oxide) (PEO) composite thin films were reported as the light-emitting layer in work by Ma and co-workers. By controlling the molar ratios of organic (benzylammonium iodide) to inorganic salts (CsI and PbI_2), the group obtained luminescent quasi-2D perovskite thin films exhibiting tunable emission colors from red to NIR. LEDs with an emission peaking at 680 nm exhibited a brightness of 1,392 cd m^{-2} and an EQE of 6.2%. The EL intensity relatively dropped by 20% following 4 h of continuous operation, and these devices exhibited enhanced spectral stability.

Kim *et al.* reported the charge carrier recombination and ion migration induced by photoexcitation by analyzing steady-state and transient PL and photoresponsivity in perovskite films [49]. These results indicate organic ligands in perovskite films efficiently prevent ion migration and ion-migration-induced defects. This efficient confinement of electron–hole pairs and prevention of ion migration in perovskite films induced high photostability and PLQY.

4.2.2 *Phosphor-converted light-emitting diodes*

MHPs with fascinating optical properties including high brightness, tunable emission wavelength, high color purity, and high defect tolerance have been regarded as promising alternative down-shifting phosphors for phosphor-converted LEDs, in which broad-band emitters are favorable for display technology while narrow-band emitters are widely used in the lighting application.

4.2.2.1 *Phosphors with narrow-band emission*

Color conversion phosphors coat blue- or UV-light-emitting diodes to yield a specific color for displays [50]. The color of a material's luminescence can be quantified by its CIE (International Commission on Illumination) coordinates. In $CsPbX_3$ NCs, layered lead halide perovskites and 0D Mn-based metal halides, the narrow, free-excitonic photoluminescence produces high-purity colors that can be tuned across the entire visible electromagnetic spectrum [51]. This leads to a wide color gamut, which is important for backlit displays to render a large range of colors. As with many of the luminescence metrics of the layered perovskites, PLQY of bulk samples exhibits a strong dependence on structure. For the blue-light-emitting Pb-Br perovskites, increased distortion of the Pb-Br lattice, as proxied by nearest-neighbor Pb-Br-Pb and Br-Pb-Br bond angles correlated with the increased quantum efficiency (22%) in $(C_6H_5(CH_2)_2NH_3)_2PbBr_4$ compared with $(C_4H_9NH_3)_2PbBr_4$ (0.36%) or $(C_6H_5CH_2NH_3)_2PbBr_4$ (3.8%). Nanostructured metal halide is a less-explored yet highly promising selection to increase quantum efficiency of these blue emitters, which are potentially attractive for wide-gamut displays. For example, the PLQY of $(C_4H_9NH_3)_2PbBr_4$ monolayers is ca. 26%. Intriguingly, exfoliating crystals of $(C_6H_5(CH_2)_2NH_3)_2PbBr_4$ is reported to result in high PLQY values of ca. 79% for blue emission. The massive differences in PLQY values obtained for the same perovskite composition appear to be a result of different synthetic methods. The Mn–Mn dependences in 0D Mn^{2+}-based metal halides were used to enhance PLQY, as shown in Figure 4.11 [52]. However, the understanding on how these synthetic methods modulate defects, surfaces, and recombination rates is limited, presenting an opportunity to closely

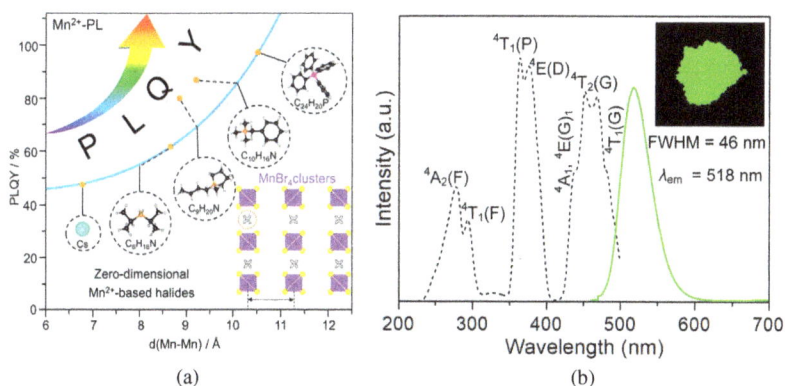

Figure 4.11. (a) Mn–Mn dependences of the emission intensity of Mn^{2+} on the closest Mn–Mn distance in 0D Mn^{2+}-based metal halides. (b) PLE and PL spectra of $(C_{10}H_{16}N)_2MnBr_4$. The inset shows the digital photograph under 365 nm UV lamp. Reproduced from Ref. [52].

study these relationships to enable the design of more efficient color-conversion phosphors. In particular, semiconductor emitters with narrow emission line widths and high PLQY values are still needed to improve LEDs relying on III–V or organic emissive layers, especially with respect to blue-light-emitting phosphors.

4.2.2.2 *Phosphors with broad-band emission*

The white-light-emitting perovskites have potential as single-source phosphors in solid-state lighting devices. In contrast to the narrow free-excitonic emission that affords high color purity (resolution), the broad emission (400–700 nm) yields high color rendition. The CRI quantifies how accurately a light source (in this case, a material's photoluminescence) portrays a target object's color. High-CRI light sources are commercially relevant, as they are pleasing to the eye, and several applications require highly accurate color representation [53]. In order to provide high color rendition, the light source must contain all colors of the visible spectrum. Owing to their broad-band emission, many of the lead halide white-light emitters exhibit CRI values above 80, a typical threshold value for commercial-quality illumination. The highest CRI reported for layered perovskite is 93 in the (110) perovskite $(C_3N_2H_4(CH_2)_3NH_3)PbCl_4$, and other high values have also been recently reported. The emission from commercial phosphors (such as Ce^{3+}-doped yttrium aluminum garnet) used for artificial illumination is not continuous across the visible spectrum, and the color rendition is improved through the use of multiple emitters. However, using a single phosphor whose emission covers the entire visible spectrum would eliminate problems associated with multiple emitters, such as self-absorption (due to overlapping absorption and emission spectra of the various components) and changes to the emission color owing to the varying degradation rates of the different components.

Variations in spectral energy and shape of the white-light-emitting perovskites also yield different "colors" of white light, affording both warm (with redder) and cold (with bluer) white light. The CCT of the emission corresponds to the temperature at which an ideal blackbody emits light of the same color; perovskite white-light emitters can be tuned to produce both warm (CCTs < 3,000 K) and cold (CCTs > 5,000 K) white

Table 4.3. PLQE, CRI, and CIE Values for an Array of White-Light-Emitting Perovskites.

Compound	PLQE	CRI	CIE
(N-MEDA)PbBr$_4$	0.5	82	(0.36, 0.41)
(EDBE)PbBr$_4$	9	84	(0.39, 0.42)
(EDBE)PbCl$_4$	2	81	(0.33, 0.39)
α-(DMEN)$_2$PbBr$_4$	Not reported	73	(0.28, 0.36)
(AEA)PbBr$_4$	Not reported	87	(0.29, 0.34)
(PEA)$_2$PbCl$_4$	<1	84	(0.37, 0.42)
(EA)$_4$Pb$_3$Cl$_{10}$	Not reported	66	(0.27, 0.39)
(API)PbCl$_4$	<1	93	(0.36, 0.37)
(MPenDA)PbBr$_4$	3.4	91	(0.24, 0.23)
(CyBMA)PbBr$_4$	1.5	Not reported	(0.23, 0.29)
(HMTA)$_4$PbMn$_{0.69}$Sn$_{0.31}$Br$_8$	73	95	(0.32, 0.33)

Notes: N-MEDA = N1-methylethane-1,2-diammonium, EDBE = 2,2'-(ethylenedioxy) bis(ethylammonium), DMEN = N1,N1-dimethylethane-1,2-diammonium, AEA = 3-(2-ammonioethyl)anilinium, PEA= phenethylammonium, EA = ethylammonium, API = 1-(3-ammoniopropyl)-1H-imidazol-3-ium, MPenDA = 2-methyl-1,5-pen-tanediammonium, CyBMA = cis-1,3-bis(ammoniomethyl)-cyclohexane, HMTA=N-benzylhexamethylenetetrammonium.

light. Halide substitution provides a straightforward method to change the CCT of the emission. For example, in (N-MEDA)PbBr$_{4-x}$Cl$_x$, increasing Cl content changes the emission CIE from a yellowish white (CCT = 4,669) to a bluish white (CCT = 6,502). The ability to easily tune the phosphor's color is an advantage for matching the CIE coordinates of current lighting sources, where the color standards have already been set (e.g., military applications).

While color control is vital for potential phosphor materials, other criteria, such as stability, are equally important. The initial studies on white-light-emitting perovskites such as (EDBE)PbBr$_4$ showed promising thermal stability over 200°C, which is higher than the typical operating temperature for LEDs. Additionally, under constant UV illumination, the perovskite's emission was unchanged after 3 months under vacuum. However, much more work remains to be done in assessing the stability of these materials, including incorporating into solid-state lighting assemblies under typical operating conditions and assessing performance, keeping in mind the typical thermal and photochemical decomposition pathways for organic–inorganic metal halides.

Despite the significant progress in understanding white-light emission in perovskites, challenges remain to be dealt with before these materials can approach commercial viability. Although films of the 2D perovskite (EDBE)PbBr$_4$ and microscale crystals of the 1D Pb-Br hybrid (H$_3$CNH$_2$(CH$_2$)$_2$NH$_2$CH$_3$)PbBr$_4$ exhibit PLQY values of ca. 9% for warm white light and ca. 12% for cold white light, respectively, these values are significantly lower than those of commercial inorganic yellow phosphors such as Ce^{3+}-doped yttrium aluminum garnet. However, unlike inorganic phosphors, films of these hybrid phosphors can be inexpensively deposited from solution and they may be suited for niche luminescence applications involving large-area coatings.

The full-spectrum white light still remains a challenge as found in the two-dimensional hybrid material (EDBE)PbBr$_4$, which exhibits the intrinsic free exciton (FE) and broad-band self-trapped exciton (STE) emission upon 365 nm ultraviolet excitation, and a combined strategy has been proposed through doping the Mn^{2+} ions enabling a super-position of multiple emission centers toward the ultra-broad- band warm white light (Figures 4.12(a) and 4.12(b)) [54]. Recently, a unique urea-amide cation has been employed to form corrugated 1D crystals by interacting with bromide atoms in lead octahedra via multiple hydrogen bonds. Further tuning the stoichiometry, cations are bonded with water molecules to create a larger space that isolates individual lead bromide octahedra. It leads to zero-dimension (0D) single crystals, which exhibit broadband "warm" white emission with photoluminescence quantum efficiency five times higher than 1D counterpart (Figures 4.12(c) and 4.12(d)] [55].

4.2.3 *Lasing*

4.2.3.1 *Photophysical process*

A detailed understanding of the outstanding optoelectronic and photonic properties exhibited by MHPs comes from a meaningful analysis of the dynamics of photoexcited states, including carrier generation,

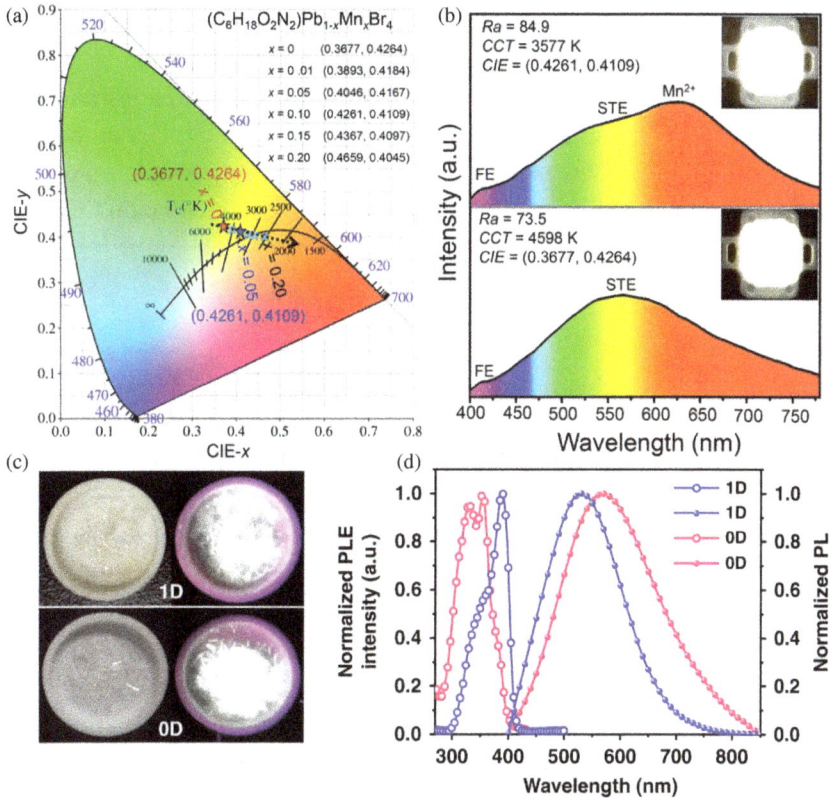

Figure 4.12. (a) CIE chromaticity coordinates of $(C_6H_{18}N_2O_2)Pb_{1-x}Mn_xBr_4$ phosphors ($x = 0, 0.01, 0.05, 0.10, 0.15, 0.20$) and the fabricated WLEDs. (b) Photographs and emission spectra of WLEDs fabricated by $(C_6H_{18}N_2O_2)PbBr_4$ and $(C_6H_{18}N_2O_2)Pb_{0.95}Mn_{0.05}Br_4$ phosphors driven by a UV chip ($\lambda_{em} = 365$ nm). Reproduced from Ref. [54]. (c) Photos of 1D and 0D crystals in ambient and under UV light (365 nm), respectively. (d) Excitation (open circle) and emission (circle) spectra of 1D (blue) and 0D (pink) crystals at room temperature. Reproduced from Ref. [55].

thermalization, cooling, scattering, and recombination. Here, we mainly discuss the recombination mechanism of photoexcited states, which directly determines the perovskite laser performances. Generally speaking, MHPs show modest exciton binding energies, 50–120, 20–100, and 2–80 meV for $ABCl_3$, $ABBr_3$, and ABI_3, respectively [56]. The large variation in experimentally determined values of exciton binding energies for each kind of perovskites results from the differences in material quality and measurement methods and inaccurate assumptions in theoretical models and calculations. Today, values in the lower range are more accepted. The modest exciton binding energies indicate that the excitons and free charge carriers coexist in perovskites at room temperature. Under thermodynamic equilibrium, the population distribution between excitons and free carriers is determined by the following Saha–Langmuir equation [57]:

$$\frac{\Phi_{fc}^2}{1-\Phi_{fc}} = \frac{1}{n}\left(\frac{2\pi m_r k_B T}{h^2}\right)^{3/2} \exp\left(-\frac{E_B}{k_B T}\right) \tag{4.20}$$

where Φ_{fc} is the fraction of free carriers, n is the total excitation density, h is the Planck constant, m_r is the reduced exciton mass, and E_B is the exciton binding energy. Figure 2.5(a) plots the calculated free charge fraction over the total excitation density at room temperature for different exciton binding energies. At low excitation densities, almost all photoexcited states are free electrons and holes, which rationalizes the efficient charge carrier extraction for PVs ($n < 10^{15}$ cm^{-3}). With the increase of excitation density, the exciton fraction increases at the expense of free charge carriers. However, the Saha–Langmuir equation becomes invalid at high carrier densities due to many-body effects. The semiconductors undergo a Mott transition from the exciton to free carrier regime at high excitation densities because the attractive force between an electron and hole is mitigated by the Coulomb Screening of other charge carriers. The Mott transition occurs at a critical carrier density (n_{crit}), usually referred to as the Mott density.

The reported Mott densities of perovskites exhibited a large variation as well, which is caused by incorrect Mott density equations [58]. The equation derivation errors mainly arise from the confusion between two different systems of units (International System of Units (SI) and Gaussian units) and the confusion between charge carrier and electron–hole pair densities. Here, we provide a derivation of the Mott density equation in both SI and Gaussian units. The critical carrier density can be estimated through the Mott criterion

$$a_B l_D^{-1} = 1.19 \tag{4.21}$$

where a_B is the exciton Bohr radius and l_D is the Debye screening length of semiconductors. According to the Wannier–Mott hydrogenic model, the exciton Bohr radius, exciton binding energy, and Debye screening length are given by

$$a_B = \frac{\varepsilon\, \hbar^2 (4\pi\varepsilon_0)}{m_r e^2} \qquad (4.22)$$

$$E_B = \frac{m_r e^4}{2\varepsilon^2\, \hbar^2 (4\pi\varepsilon_0)^2} \qquad (4.23)$$

$$l_D = \sqrt{\frac{\varepsilon\, k_B T (4\pi\varepsilon_0)}{4\pi\, n e^2}} \qquad (4.24)$$

where $\hbar$ is the reduced Planck constant, ε is the relative dielectric constant, ε_0 is the vacuum permittivity, m_r is the reduced mass of the electron and hole, e is the elementary charge, k_B is the Boltzmann constant, T is the temperature, and n is the charge carrier concentration. Noteworthily, the equations with and without the $(4\pi\varepsilon_0)$ correspond to SI and Gaussian units, respectively. In both systems of units, the same Mott density equation can be obtained as follows:

$$n_M = \frac{1.19^2\, k_B T}{8\pi\, a_B^3 E_b} \qquad (4.25)$$

The Mott density of MAPbI$_3$ is estimated to be ~4×10^{17} cm^{-3}. The free carriers dominate in MAPbI$_3$ even below the Mott density due to small exciton binding energy (only a few millielectronvolts) [59]. Thus, we can argue that the lasing originates from bimolecular electron–hole recombination from a theoretical point of view, though the exact lasing mechanism needs further experimental investigations. The situation becomes complicated for CsPbBr$_3$, for which the Mott density is ~2×10^{17} cm^{-3}. The reported threshold carrier densities of CsPbBr$_3$ lasers just locate in the vicinity of the Mott density ranging from 0.7×10^{17} to 6×10^{17} cm^{-3}, where the excitons and free carriers share similar population as a result of a modest exciton binding energy. It is difficult to attribute the lasing to a certain recombination mechanism according to the population distribution between the exciton and free carrier. Note that a blue-shift of the lasing modes with excitation density is ubiquitous in perovskite lasers. The mode blue-shift is a typical characteristic of lasing in the electron–hole plasma regime.

Therefore, the electron–hole plasma recombination is more likely to be responsible for the stimulated emission in MHPs. Very recently, Schlaus *et al.* provided a clear picture of the lasing mechanism in $CsPbBr_3$ nanowires. The laser dynamics in $CsPbBr_3$ nanowires was probed using ultrafast spectroscopy (Figure 4.13(a)) [60]. Upon excitation, a short-lived broad lasing spectrum emerges first, which can be attributed to hot carrier recombination. Then the spectral narrowing and bathochromic-shift occur owing to the hot carrier cooling process. As the carrier density decreases significantly, the optical gain profile blue-shifts toward the peak energy of spontaneous emission. This signifies that the stimulated emission of electron–hole plasma involves the emission of a plasmon quanta. The plasmon quanta is a collective oscillation of electron–hole plasma with frequency proportional to the square root of carrier density. In a word, the lasing in $CsPbBr_3$ nanowires originates from the stimulated emission of electron–hole plasma coupled with plasmon emission (Figure 4.13(b)) [60]. With larger exciton binding energies, the chloride perovskites probably support excitonic lasing, but this remains

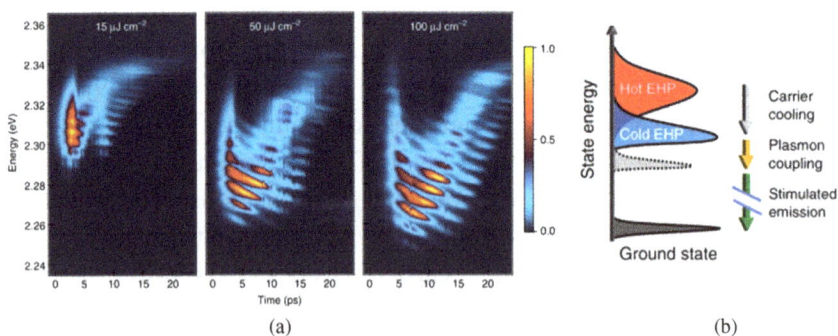

Figure 4.13. (a) Time-resolved lasing spectrum of $CsPbBr_3$ nanowires. (b) Schematic of stimulated emission from electron–hole plasma coupled with plasmon emission. Reproduced from Ref. [60].

to be verified through experiments. The low-dimensional perovskite materials, such as perovskite QDs and 2D RP perovskites, have large exciton binding energies due to the quantum confinement. Size-dependent bandgaps manifested the quantum confinement effect in perovskite nanocrystals. The lasing actions have been reported in many kinds of perovskite nanocrystal films. Different gain mechanisms were put forward to account for the lasing actions in perovskite nanocrystals, including exciton, biexciton, and trion gains. To date, the lasing actions are mainly realized in perovskite nanocrystals with weak quantum confinement, which would not significantly increase the binding energies of the exciton, biexciton, and trion. The exact lasing mechanism in perovskite nanocrystal films remains open to question. The 2D RP perovskites have multiple quantum well structures, exhibiting obvious perovskite layer number-dependent exciton absorption and emission. In such RP perovskites, excitonic lasing has been demonstrated unambiguously.

The charge carriers in MHPs exist as large polarons. A polaron is a quasiparticle consisting of an electron or hole dressed by its self-induced lattice polarization in polar semiconductors [61]. There are small and large polarons, formed through two fundamental interaction potentials, i.e., long-range Coulomb potential and short-range deformation potential, respectively. The polarons in MHPs are proved as large polarons formed predominantly from the deformation of lead halide sublattice $[PbX_6]^{4-}$ frameworks. A direct consequence of large polaron formation is the efficient screening of the Coulomb potential. All events requiring the Coulomb potential, such as scattering of a charge carrier with charged defects or other charge carriers, are inhibited. Accordingly, the trap-assisted non-radiative recombination and many-body Auger recombination are suppressed effectively in MHPs. The protection of charge carriers as large polarons results in their excellent optoelectronic and photonic properties, such as ambipolar carrier transport and high-efficiency luminescence.

4.2.3.2 *Optical amplification using perovskites*

Stimulated emission is the coherent photon emission process wherein an incident photon is amplified (gain) via downward radiative transitions from excited state levels. By studying charge transfer process in perovskite solar cells, researchers first observed stimulated emission in MAPbX$_3$ with the aid of ultrafast pump–probe spectroscopy [62].

Optical amplification, measured via gain per unit length (in cm^{-1}), can be studied by measuring the absorption spectra from the photoexcited gain medium. In perovskite thin films, optical gains have been reported in the range 3,200 ± 800 cm^{-1} [63]. This value is comparable to that of single-crystal GaAs and agrees with the known high absorption coefficient of perovskite. Gain has been shown to be as long as 200 ps, with a threshold of ~16 μJ cm^{-2}. When one varies the pump fluence, the emitted spectral intensity yields a threshold for amplified spontaneous emission (ASE). The luminescence becomes narrow and sharply increases above the ASE threshold. Reported ASE thresholds for halide perovskites range from 12 μJ cm^{-2} under femtosecond excitation to 60 μJ cm^{-2} under nanosecond photoexcitation conditions and 7.6 μJ cm^{-2} at pulse durations as long as 5 ns with a liquid crystal reflector.

4.2.3.3   *Original reports of perovskite lasing*

Solution-processed MAPbX$_3$ perovskite films were studied with an eye to lasing by Xing *et al.* [64]. An increased pump fluence leads to the transition from spontaneous emission (SE) to ASE in MAPbI$_3$ thin films. Using the threshold fluence (12 ± 2 μJ cm^{-2}) and absorption coefficient ($\alpha = 5.7 \times 10^4$ cm^{-1} at 600 nm), the ASE threshold was calculated to be ~1.7 $\times$ 10^{18} cm^{-3}. The Auger recombination process in perovskites, dominant at high pump fluence, leads to lifetimes in the range of a few ps to ns, depending on the photogenerated charge carrier density. The low ASE threshold in perovskites is attributed in significant part to the low trap density in the best of such films.

The PLQY as a function of excitation fluence in CH$_3$NH$_3$PbI$_{3-x}$Cl$_x$ perovskite films was, as previously discussed, an impressive 70% at high excitation density. Friend and co-workers constructed and demonstrated the operation of vertical cavity lasers that employed a perovskite layer sandwiched between a dielectric mirror and a metal mirror.

The relation between the phase-transition of the perovskite film and ASE mechanisms was investigated by studying behavior in different phase states. A sharp ASE peak with the maximum spectral intensity occurs at 120 K. PL emission spectra show a notable variation of the spectral characteristics when the temperatures move from 120 to 160 K, in which range the structural phase transformation from orthorhombic to tetragonal occurs. As the temperature exceeds 120 K, the lasing threshold continues to increase because of thermal broadening of the gain peak. Low-threshold ASE and lasing has also been reported from colloidal $CsPbX_3$ NCs with a 10 nm size. Optical amplification in perovskite NCs was obtained with low pump thresholds (5 ± 1 mJ/cm^2) and across the visible spectral range (440–700 nm). Conformally coated perovskite NCs on a high-finesse resonator, such as silica microspheres, have enabled the observation of whispering-gallery-mode lasing.

4.2.3.4 *Laser cavity structures in perovskite lasers*

In semiconducting nanowires, waveguiding is provided along the axial direction, and the two end facets form a Fabry–Perot cavity. The growth of high-quality single-crystal nanowires from low-temperature solution-processing of $MAPbX_3$ perovskites was reported by Zhu and co-workers. The perovskite nanowires exhibited low lasing thresholds (200 nJ/cm^{-2}), high quality factors ($Q \sim 3,600$), near-unity QY, and wavelength tunability from the NIR to the visible wavelengths [65].

Perovskite nanoplatelets were grown on mica substrates and formed whispering gallery mode (WGM) cavities as a result of the growth procedure. Individual perovskite nanoplatelets were photoexcited using a femtosecond-pulsed laser, and a lasing threshold of 37 μJ cm^{-2} was achieved in $MAPbI_3$ nanoplatelets. Solution growth of $MAPbBr_3$ perovskite microdisks (also a means to WGM microresonators) exhibited a quality factor of ≈ 430.243 single-mode lasing at 558 nm was achieved in a $2 \times 2 \times 0.6 \ \mu$m^3 square microdisk and exhibited a threshold of $3.6 \pm 0.5 \ \mu$J cm^{-2}. WGM lasers from planar perovskite nanoplatelets exhibit tunable optical modes and impressive optical gain and offer a pathway to integration with Si technologies.

In distributed feedback (DFB) cavity structures, optical feedback is provided via Bragg scattering from an interference grating built either directly into the active medium or in the vicinity of the resonator through periodic alternation of the refractive index. Jia *et al.* reported a metal-clad, second-order DFB perovskite laser with a threshold of 5 kW/cm^2 for durations $\leq$25 ns. The work demonstrated that the substrate can be used to dissipate heat and offered thus a step in the direction of future electrically driven architectures. Saliba *et al.* reported perovskite DFB cavities made by nanoimprinting a corrugated structure onto a polymer template, after which they then introduced a conformal perovskite layer. Significant narrowing of the line width of the ASE peak was observed with increased excitation fluence, and the wavelength red-shifted as the grating periodicity of the DFB cavity increased. The fluence threshold diminished from 2.1 to 0.3 μJ cm^{-2} as the grating period decreased from 420 to 400 nm.

A new hybrid perovskite vertical cavity surface-emitting laser (VCSEL) was microfabricated based on a uniform perovskite thin ($\approx$300 nm) film placed between two high-reflectivity ($\approx$99.5%) distributed Bragg reflectors (DBRs) [66]. This work leveraged gallium nitride semiconductor process innovations wherein nanoporous-GaN (NP-GaN) provides the low-index layers in a NP-GaN/GaN (electrically conducting) multilayer stack. The perovskite VCSEL device enabled the study of gain dynamics on excitation time scales that are much longer than typically reported in femtosecond experiments in the perovskite literature. VCSEL lasing output as a function of pumping fluence indicated a lasing threshold of 7.6 μJ cm^{-2}. Continuous-wave (CW) operation of a perovskite methyl-ammonium lead iodide (MAPbI$_3$) laser was recently reported. Under constant blue optical photo excitation, devices emitted near-infrared (785 nm wavelength) laser light over the course of an hour when operated at 100 K. To achieve CW lasing, a strategy to generate exciton/charge-trapping sites in the material was sought by taking advantage of the phase transition in MAPbI$_3$. When the temperature is decreased to 160 K, the crystal structure of MAPbI$_3$ changes from the tetragonal phase to the orthorhombic phase. A DFB laser fabricated with MAPbI$_3$ was operated at 100 K, hence in orthorhombic phase. With a high excitation power (>17 kW cm^{-2}), the tetragonal phase can transform to the orthorhombic lattice within 100 ns, due to local heating, and this provides a local heterostructure inside the perovskite film.

4.3 Photodetectors

In addition to their use in PVs and LEDs, halide perovskites have recently emerged as promising candidates for new-generation photodetectors. However, conventional three-dimensional halide perovskite photodetectors suffer from current–voltage hysteresis, unreliable performance, and instability. By controlling the morphological dimensionality of bulk three-dimensional and quasi-two-dimensional structured perovskites, low-dimensional halide perovskites (LDHPs, such as two-dimensional, one-dimensional, and zero-dimensional halide perovskites), which have robust chemical stability and optoelectronic tunability owing to the quantum size effect, can be obtained, providing an alternative solution to overcome the aforementioned limitations.

4.3.1 *Mechanisms and Architectures of Photodetectors*

This section summarizes and discusses the operation mechanisms and photodetection performance of widely investigated perovskite-based photodetectors. Figure 4.27 shows that the two-terminal devices include photodiodes and photoconductors, and the three-terminal device refers to phototransistors with source, drain, and gate electrodes. According to the spatial layout of the photoactive medium and electrodes, photodiodes and vertical photoconductors have a small electrode spacing with a short carrier transit length, which together provide fast response, as well as low driving voltage; in contrast, lateral photoconductors and phototransistors lead to slow response and high driving voltage due to their large electrode spacing, where the response speed is sacrificed to maintain a high photocurrent. However, lateral devices have additional advantages that are typically much more reproducible and easy to produce. Generally, in photodiodes, an absorbed photon can generate at best one electron–hole pair with an EQE of no more than 100%, which requires non-injecting metal-semiconductor contacts. Usually, the introduction of interface layers (hole- or EBLs) cannot guarantee an increase in responsivity, but can effectively reduce the dark current under reverse bias to obtain high detectivity, accompanied by small NEP and large LDR. In contrast, when metal-semiconductor contacts allow at least one type of carrier injection

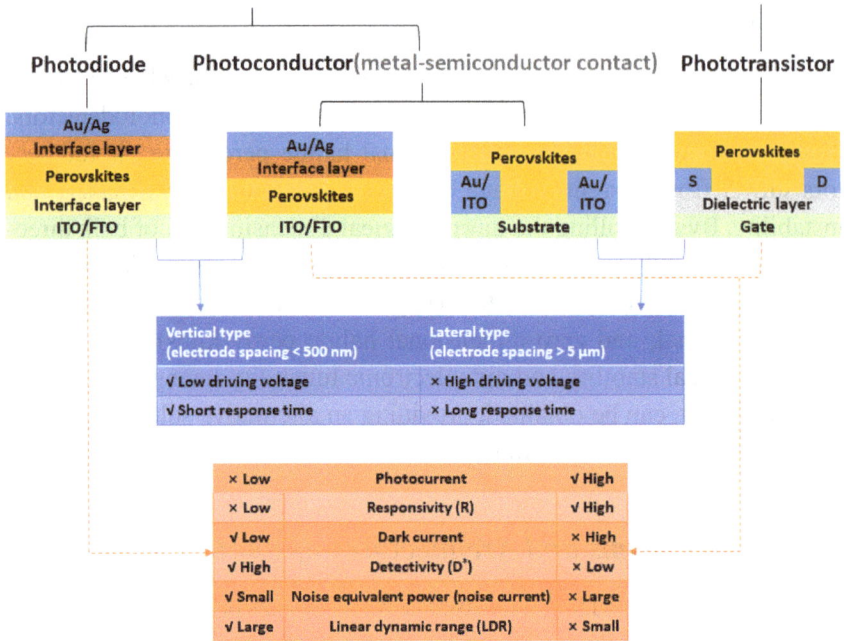

Figure 4.14. Schematic of architectures for perovskite-based photodetectors and the characteristics of photodetection performance for different devices.

into the device under illumination, EQE in photoconductors and phototransistors can exceed 100%, which results in high photocurrent, and consequently, high responsivity. Due to the lack of dark current suppression, the performances of detectivity, NEP, and LDR must be sacrificed.

Figure 4.14 and the following section summarize and compare the important figure-of-merit parameters of the above photodetector configurations based on the versatile compositions, structures, and morphologies of perovskite materials.

4.3.2 *Photodetector Performance Metrics*

The performance metrics used to quantitatively characterize photodetectors include the on–off ratio, EQE, responsivity, specific detectivity, linear dynamic range (LDR), and temporal response, as summarized in what follows.

4.3.2.1 *On–off ratio*

The on–off ratio is defined as the ratio of photocurrent (I_{light}) to the dark current (I_{dark}). To determine the on–off ratio, the incident power, wavelength, and bias voltage must be taken into account.

4.3.2.2 *Responsivity*

The responsivity I is a function of the wavelength of the incident light and characterizes the electrical output per optical input. A simple expression for R is as follows:

$$R = \frac{J_{photo}}{P_{light}} = \frac{J_{light} - J_{dark}}{P_{light}} \tag{4.26}$$

where J_{photo} is the photocurrent density, J_{light} is the current density under light, J_{dark} is the dark current density, and P_{light} is the incident light intensity.

4.3.2.3 *External quantum efficiency*

The external quantum efficiency (*EQE*, also called the spectral response) is the ratio of the number of charge carriers collected by the photodetector at a fixed bias to the number of photons of a given light beam; therefore, it depends on both the absorption of light and the collection of charges. In some cases, it can exceed 100%. It is calculated as follows:

$$EQE = \frac{Rh\upsilon}{e} \tag{4.27}$$

where $h\upsilon$ is the energy of the incident photons and e is the absolute value of the electron charge.

4.3.2.4 *Specific detectivity*

Specific detectivity is used to characterize the weakest light that can be detected by the photodetector. Specific detectivity is given by the following expression:

$$D* = \frac{\sqrt{A\Delta f}}{\text{NEP}} \tag{4.28}$$

where A is the active area of the device, Δf is the electrical frequency bandwidth, and NEP is the noise-equivalent power, which is defined as the incident power at which a signal can no longer be extracted from the noise current. When the device noise is dominated by shot noise, $D*$ can be simplified as

$$D* = \frac{R}{\sqrt{2eJ_{\text{dark}}}} \tag{4.29}$$

Furthermore, according to Eqs. (4.28) and (4.29), the specific detectivity of the devices can be calculated based on noise measurement analysis and dark current measurements. For the latter, as previous works suggested, if the shot noise from the dark current is the dominant factor and the thermal and flicker noise contributions are negligible, then the specific detectivity can be expressed as Eq. (4.29). However, this estimation of the specific detectivity may not be appropriate and could lead to overestimation when ignoring the thermal and flicker noise contributions [67].

4.3.2.5 *Light-dependent resistor*

The light-dependent resistor? (LDR) corresponds to the range over which the photocurrent is proportional to the optical power. Although the definition of the LDR is simple, one can find many discrepancies in the methodologies reported in the literature. Generally, the LDR can be expressed as

$$\text{LDR} = B \log \frac{J^*_{ph}}{J_d} \tag{4.30}$$

where B is a constant and J^*_{ph} and J_d are the maximum and minimum detectable current densities, respectively, in which J_d is limited by the noise current of the photodetector. According to electrical engineering reports, the constant B is either 20 (which is conventionally used to convert the ratio between the maximum and minimum voltages to decibels) or 10 (which is used to convert the ratio between the maximum and minimum currents to decibels) [68]. This issue can create confusion for researchers, for example, when a proposed device with an LDR based on Eq. (4.30) with $B = 20$ is compared to a photodetector with an LDR calculated with $B = 10$.

4.3.2.6 *Temporal response*

The temporal response is evaluated based on the rise time t_r (time in which the device response increases from 10% to 90%) and the fall time t_f (time in which the device response falls from 90% to 10%). This parameter is related to the trapping and detrapping of the photogenerated charges and to the quality of the contact between different layers in the device. Additionally, the temporal response should be optimized based on the thickness and mobility of the active layer and the voltage applied to the device.

4.3.3 *Different Materials Systems for Photodetectors*

Polycrystalline films and single crystals are the two main forms of perovskites. For film formation, a fine control over the crystallization process is critical to engineer the film morphology, such as the uniformity and surface coverage, which essentially determine the device performance. Because photodetectors operate under a similar principle to that of solar cells that absorb incident photons to produce free electrons and holes, contributing to the electrical current, the morphological optimization avenues for the preparation of high-quality perovskite films with less defects (pinholes and GBs) can follow the relatively developed techniques for solar cells. Compared to perovskite polycrystalline films suffering from morphological disorder, perovskite single crystals exhibit the advantages of low trap density, low intrinsic carrier concentration, high mobility, and long diffusion length, which make the material ideal for simultaneously realizing rapid and sensitive photodetection. From 3D large-sized crystals, 2D nanosheets, 1D nanowires, to 0D QDs, versatile perovskite single crystals with excellent structural and photoelectric properties hold huge potential for high-performance photodetectors. Next, according to the morphological category of perovskites used in photodetectors, we turn to summarize their preparation methodologies and spectral response characteristics.

4.3.3.1 *Polycrystalline films*

Some universal strategies, such as thermal annealing, solvent annealing, atmospheric control, solvent engineering, chemical vapor deposition, vacuum flash-assisted solution processing, and the use of additives, have been successfully employed to yield high-quality perovskite films for PV applications [16]. These strategies are also generally applicable to perovskite film photodetectors. With respect to solar cells, the target of photodetectors is the delivery of a photocurrent signal, rather than that of electric power, to a load. Thus, uniform and dense perovskite layers with less pinholes and GBs benefit the charge photo-generation, transport, and/or collection.

4.3.3.2 *Bulk crystals*

The currently reported methods for the growth of bulk perovskite crystals used in photodetectors include inverse temperature crystallization, antisolvent vapor-assisted crystallization, and gradually cooling-induced crystallization techniques. The inverse temperature crystallization method is suitable for those materials whose solubility in particular solvents is high at room temperature, but decreases with increasing temperature. $CH_3NH_3PbX_3$ perovskites exhibit inverse temperature solubility behavior in g-butyrolactone, DMF, DMSO, or their mixture. This phenomenon can be used to rapidly crystallize $CH_3NH_3PbX_3$ in hot solutions, as illustrated in Figure 4.15(a) [69]. Figure 4.15(b), which shows the solubility of $CH_3NH_3PbCl_3$ powder in a mixture of DMSO-DMF (1 : 1 v/v), clearly demonstrates the decreased solubility at elevated temperatures. A highly transparent and colorless $CH_3NH_3PbCl_3$ crystal can be obtained with parallelepiped shape. The same method was used to grow $CH(NH_2)_2PbI_3$ and $CsPbBr_3$ crystals, respectively, for photon detection. The antisolvent vapor-assisted crystallization method uses a solvent with high solubility and moderate coordination for CH_3NH_3X and PbX_2 (DMF or g-butyrolactone) and an antisolvent in which both perovskite precursors are completely

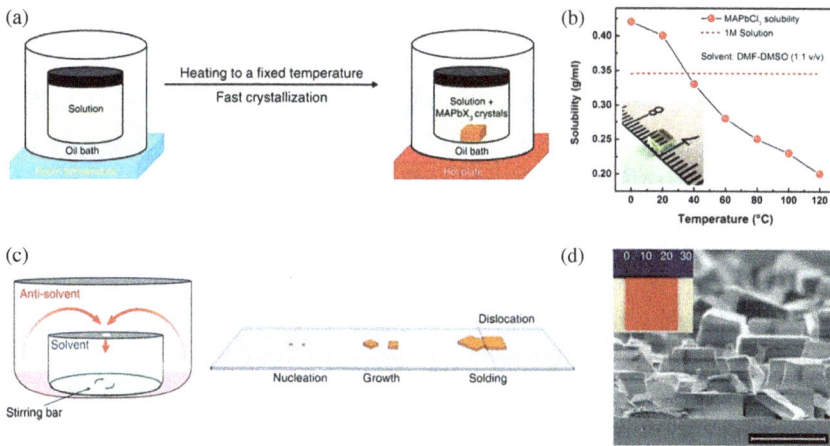

Figure 4.15. (a) Schematic of the inverse temperature crystallization process, in which the solution is heated from room temperature and kept at an elevated temperature to initiate the crystallization. (b) Solubility data for MAPbCl$_3$ powder in DMSO-DMF (1 : 1 v/v) at different temperatures. Inset: Transparent single crystal grown from MAPbCl$_3$ solution. (c) Schematic of the antisolvent vapor-assisted crystallization process, in which an appropriate antisolvent is slowly diffused into the saturated solution of the crystal precursors, leading to the crystallization, and the kinetics of perovskite crystal growth. (d) Cross-sectional SEM of laterally continuous perovskite crystals on the substrate. Reproduced with permission from Ref. [70].

insoluble (dichloromethane). The antisolvent is then slowly diffused into the solution of precursors, leading to the growth of perovskite crystals, as shown in Figure 4.15(c). Saidaminov *et al.* succeeded in growing CH$_3$NH$_3$PbBr$_3$ crystals on a planar substrate and over multiple cm^2 areas, as shown in Figure 4.15(d), whose integration was further exploited to construct high-performance photodetectors [70]. Walters *et al.* obtained large-sized CH$_3$NH$_3$PbBr$_3$ single crystals by using this method and fabricated the first perovskite two-photon photodetector to demonstrate the prospects of perovskites for applications in nonlinear optics.

4.3.3.3 *Nano-sheet/plate crystals/nanowire crystals*

Apart from 3D large-sized crystals, 2D/1D perovskite crystals, such as nanosheets and nanoplates, have recently shown excellent photoresponse properties. Currently, the preparation strategy of nano-sheet/plate perovskite crystals used in photodetectors mainly includes solution-phase growth and a combined solution process and vapor-phase conversion method. Song *et al.* reported for the first time the fabrication of atomically thin 2D $CsPbBr_3$ nanosheets and their application in solution-processed flexible photodetectors, with high performance. These perovskite nanosheets were synthesized through the reaction of cesium stearate (CsSt) and $PbBr_2$ precursors like the colloidal synthesis. The nanosheet edge could reach a length of more than 1.5 mm while the thickness is only about 3.3 nm [71]. Halide perovskite nanowires reported in the photodetection applications are largely based on the solution-phase synthesis method. Deng *et al.* used the one-step self-assembly method to grow $CH_3NH_3PbI_3$ nanowires, and further applied an evaporation-induced self-assembly to align single-crystalline perovskite nanowires [72]. Due to the slow crystallization process and moisture-isolated growth environment, the perovskite nanowires showed smooth surfaces and photodetectors made from the perovskite nanowire arrays exhibited extremely high responsivity. In addition, all-inorganic perovskite $CsPbX_3$ nanowires were synthesized via the standard Schlenk method in solution phase and the as-prepared nanowires could then be dropped onto the substrate for the prototypical photodetector device fabrication.

4.3.3.4 *Quantum dots and nanocrystals*

Halide PQDs and PNCs are expected to exhibit favorably integrated properties of quantum-size effects, enhancing their optical properties compared with their bulk counterparts, versatile surface chemistry, and a "free" colloidal state, allowing their dispersion into various solvents and matrices, and eventual incorporation into various devices. Alteration of the structure of monovalent cations A, divalent cations B, and monovalent halide anions X can influence the bandgap of perovskites. For example, the bandgaps of 0D PNCs/PQDs can be easily tuned by controlling the composition of the cations (MA, FA, and Cs) and halide anions (Br, Cl, and I). An ultrasonic-assisted method for synthesizing $APbX_3$ perovskite NCs (A = MA, FA, or

Cs; X = Cl, Br, or I) was reported, which produced NCs with an average side length of 10 nm and a thickness of less than 5 nm. The photoconductive detectors demonstrated a responsivity of 0.15×10^{-3} A/W at 365 nm (60 mW/cm^2) with a 2 V voltage bias. Inorganic perovskite CsPbX$_3$ (X = Cl, Br, or I) NCs have attracted considerable attention owing to their defect-tolerant nature, which is expected to lead to higher device reliability/ stability compared with hybrid inorganic–organic halide PNCs. Ramasamy *et al.* first presented facile halide exchange reactions using lithium salts (LiX, X = I, Cl, or Br) in CsPbBr$_3$ perovskite colloidal PNCs at room temperature and their application in lateral thin-film photoconductive detectors. The CsPbI$_3$ PNC-based photodetectors showed an on-off ratio of 10^5 and rise and decay times of 24 and 29 ms, respectively.

To further improve the performance of 0D-perovskite-based photodetectors, such as increasing their photoresponsivity, broadening their absorption range, and enhancing their stability, the integration of several materials with PQDs/PNCs has been proposed. Hybrid phototransistors based on CsPbBr$_3$ and PbS QDs were fabricated by Yu and co-workers. Through the introduction of a hybrid device structure and the design of the energy band of the materials, a responsivity of up to 4.5×10^5 A/W and a specific detectivity of 7×10^{13} Jones were obtained for the devices. In addition, CsPbBr$_3$-PbS QD phototransistors exhibited ultrabroadband photodetection between 400 and 1,500 nm.

4.4 X-ray Detectors

X-ray detection is important for medical diagnostics, railway flaw detection, security screening, non-destructive testing in industrial products, and quality inspection in food industry. Recently, as one of the alternative materials, halide perovskites have shown great potential in high-performance X-ray detectors for their relatively high atomic number, superior carrier life-time production, tunable band gap, and low temperature fabrication process. Accordingly, based on halide perovskite, some advanced and meaningful work was conducted, and corresponding performances of the X-ray detector have reached a comparative level.

4.4.1 *Classification of X-ray Detectors*

Typically, different-ray detectors operate with two detection modes. The dominant one is indirect X-ray detection where phosphors or scintillators have been utilized to transform X-rays into visible light that is further collected by a charge-coupled device (CCD) or photodiode for converting it to electric charge. This indirect X-ray detection has the advantages of convenience, low-cost, abundant options, and flexible conversion rate, and has been the mainstream X-ray detecting method. Besides, it shows more potential of combining with mature sensor arrays. However, the indirect detector shows a limited spatial and energy resolution because of lateral spread of light-scattering crosstalk in the converting layer, and low conversion efficiency. Additionally, the optical coupling between scintillator and thin film transistors/complementary metal-oxide semiconductor (TFT/CMOS) is weak in indirect imaging devices (for scintillator with a thickness of 200 lm, 5–10% transmission loss), which influences the imaging quality. The other detection method is direct X-ray detection where the radiation is converted into electrical currents directly and the samples are usually semiconductor films. This simple converting mode endows direct X-ray detectors with a wide linear response range, fast pulse rise time, high-energy resolution, and high spatial resolution. In this section, we mainly focus on direct detectors since they can overcome some limitations of the scintillators (optical crosstalk and non-proportionality).

The performance of X-ray detectors is largely limited by the properties of the material. Large stopping power ability (density 5 g/cm^3 and atomic number >40) is needed to effectively absorb the X-ray in a smaller thickness, suitable band gap (1.5 eV–2.5 eV) to generate more carriers under a certain X-ray energy (Figure 4.16), large mobility-lifetime products to ensure effective collection, and large resistance (beyond 10^9 cm) to reduce the dark current [73].

4.4.2 *Advantages of Metal Halide Perovskites for X-ray Detectors*

The stopping power is defined as the capability of completely absorbing X-ray. Generally, the stronger the stopping power ability of materials is,

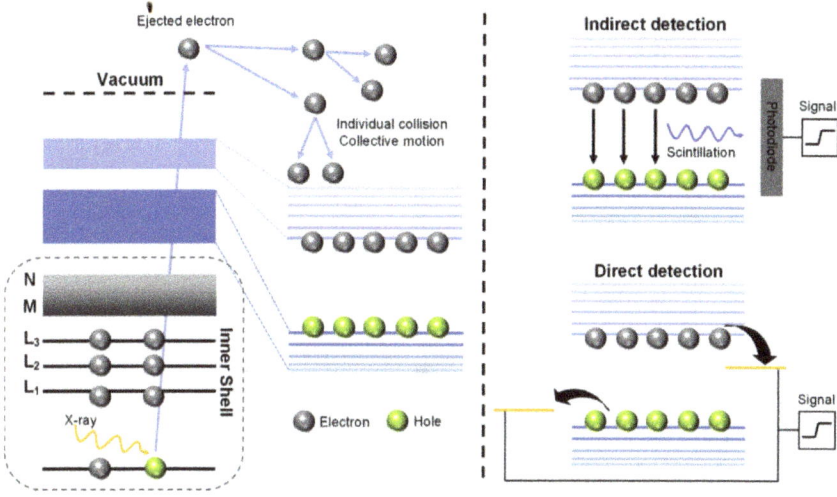

Figure 4.16. The mechanism of electron–hole generation in a semiconductor under X-ray irradiation and detection mechanism of ID and DD. Electron at the inner shell ejects after absorbing X-rays. The semiconductor releases free electrons due to the ionization of ejected electrons. The free electrons then relax to the CB, while holes migrate to the top of the VB. Here, the generated electron–hole pairs can be collected under electric field or radiatively recombine to emit visible photons. Reproduced from Ref. [73].

the smaller the material's thickness required in a certain X-ray energy. Under 50 keV X-ray photons, the thickness required for general perovskites is much smaller than that of traditional material (a-Se). The X-rays attenuation formula is used to predict the stopping ability of X-rays, listed as follows:

$$I = I_0 \exp(-\mu_m d) = I_0 \exp(-\alpha \rho d) \tag{4.31}$$

where μ_m, q, d, and α represent the linear absorption coefficient, density, thickness, and the X-ray absorption cross-section, respectively. Increasing one of the parameters in formula (4.31) can lead to a higher attenuation.

Notably, α can be explained by the following formula:

$$\alpha \propto \frac{Z^4}{AE^3} \tag{4.32}$$

where Z is the atomic number of materials, A the atomic mass, and E the X-ray photon energy. The ideal materials in direct detection should

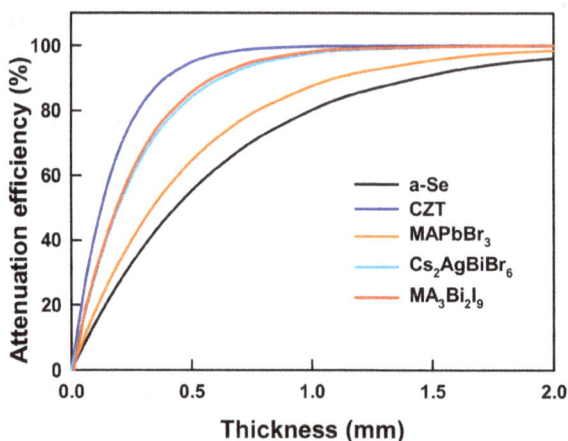

Figure 4.17.   Calculated attenuation efficiencies of a-Se, $Cd_{0.9}Zn_{0.1}Te$ (CZT), $MAPbBr_3$, $Cs_2AgBiBr_6$, and $MA_3Bi_2I_9$ semiconductors to 50 keV X-ray photons versus thickness. Reproduced from Ref. [75].

possess a large Z (>40) for completely absorbing the X-ray and turning it into electronic signals [74]. This is in accord with perovskites for the atomic numbers of elements (Pb, I, Br, Cs, Ag, etc.), which makes it moderately larger than some traditional materials (such as Si ($Z = 14$) and a-Se ($Z = 34$)). The greater the composition of I ion in it, the greater the Z value (I_3 higher than $I_{1-x}Br_x$, $0 < X < 1$). Different materials with different Z values and their attenuation efficiency are shown in Figure 4.17. Under 50 keV X-ray photons irradiation, the required thickness of the perovskite for completely absorbing X-ray is between traditional materials of CZT and a-Se, indicating that halide perovskites have the potential for X-ray detector applications [75]. In short, moderate density and high Z element make perovskites a promising candidate for X-ray detection.

4.4.3 *Direct X-ray Detection*

To qualify as a superior performance detector, a device should possess high sensitivity, low noise/noise current, low detection dose rate, quick response speed, and high carrier collection efficiency at each electrode. To be applied as an X-ray imager, it should possess high contrast and spatial resolution for clear imaging.

4.4.3.1 *Performance index of X-ray detection*

Sensitivity: High sensitivity is a key parameter for X-ray detectors, especially in medicine treatment, because higher sensitivity can shorten the exposure time of human body under radiation, thus reducing the risk of ionizing radiation. Additionally, high sensitivity indicates that large X-ray-induced current is generated under a certain X-ray dose rate, which greatly relates to the quality of imaging. Typically, sensitivity is defined as the collected charge per unit area under the exposure of radiation. Equation (4.33) is used to evaluate sensitivity [83]:

$$R_s = \frac{I_p - I_d}{D \times A} \qquad (4.33)$$

where R_s is the sensitivity, I_p, induced photocurrent, I_d, dark current, D, irradiation dose rate, and A, the effective area of the X-ray detector. Therefore, in order to pursue a high sensitivity, the irradiation dose rate and I_d should be small enough while I_p should be large. Increasing the electric field can also improve the sensitivity by reducing carrier recombination (increasing I_p). For MAPbBr$_3$-based detector, the electric field increases from 25 V cm^{-1} to 50 V cm^{-1}, and the sensitivity of X-ray detector improves from 467 μC Gy$_{air}^{-1}$ cm^{-2} to 525 μC Gy$_{air}^{-1}$ cm^{-2}, which is due to the effective carrier transport and few charge trapping. However, a large electric field will induce I_d, which damages the sensitivity, so a suitable electric field is required for better device performance.

Noise: The intensity of noise is greatly related to the sensitivity of the detector because noise is the lowest level of detectable signal, and usually stems from the intrinsic uncertainties or fluctuations. Usually, noise can induce noise current (i_{noise}) influencing the photoelectric device, which consists of four different types of noise (shot noise (i_{shot}), thermal noise ($i_{thermal}$), flicker noise ($i_{1/f}$), and generation-recombination noise (i_{g-r})). The corresponding formulas are as follows [76]:

$$i_{noise} = \left[i_{shot}^2 + i_{thermal}^2 + i_{1/f}^2 + i_{g-r}^2 \right]^{1/2} \qquad (4.34)$$

$$i_{shot} = \sqrt{2ei_d B} \qquad (4.35)$$

$$i_{thermal} = \sqrt{\frac{4kTB}{R_{sh}}} \tag{4.36}$$

$$i_{1/f} = i(f,B)_{\frac{1}{f}}^2 \tag{4.37}$$

$$i_{g-r} = i(f, B)_{g-r} \tag{4.38}$$

where e represents the elementary charge, B, the bandwidth, T, the absolute temperature, k, the Boltzmann constant, R_{sh}, the shunt resistance of the detectors, and f, the frequency, respectively. The shot and thermal noise are "white noise", which is frequency-independent. Differently, flicker noise and generation-recombination noise depend on the frequency. Besides, Pan *et al.* reported that large resistivity can lead to a small shot noise, and large band gap induces a low thermal noise. This is the reason why researchers are devoted to increasing resistance and modifyimg the band gap. Typically, the acceptable limit of dark current value in perovskite is 10 nA/cm^2, which is smaller than the requirement of a commercial detector (<50 nA/cm^2).

Signal to noise ratio (SNR): Noise not only influences the sensitivity of a detector by noise current, but also determines the lowest detectable dose rate (LoD). LoD is the foundation to evaluate the detection performance because it defines the range of potential application. For X-ray diagnostics, the LoD is 5.5 μCGy$_{air}$ s^{-1}, while radiotherapy needs high values (total dose up to 5 Gy). The International Union of Pure Applied Chemistry (IUPAC) defines the detection limitation as that where the producing signal value is three times the level of the noise, so the researchers used a SNR value of three to define the LoD in X-ray detector. The formula of SNR is as follows [75]:

$$\mathrm{SNR} = \frac{J_s}{J_n} \tag{4.39}$$

$$J_s = J_p - J_d \tag{4.40}$$

$$J_n = \sqrt{\frac{1}{N}\sum_i^N (J_i J_p)^2} \tag{4.41}$$

where J_s represents the signal current density, J_n, the noise current density, J_p and J_n, average photocurrent density and the average dark current density, respectively. Two factors greatly related to SNR are injected carriers from each contact interface and intrinsic thermally activated carriers.

Response speed: Response speed is defined as the time taken for the detector to respond to an external stimulus. It is significant in X-ray detector because fast response speed can shorten the exposure time, and enable higher frame rate during imaging. Usually, the –3 dB bandwidth is used to define the temporal response of a detector. The –3 dB bandwidth (f_{-3dB}) of photodetectors is defined as the input light modulation frequency, at which the output signal is –3 dB lower than the signal obtained under continuous wave illumination. The formula is as follows:

$$f_{-3dB}^{-2} = \left(\frac{3.5}{2\pi\tau_{tr}} \right)^{-2} + \left(\frac{1}{2\pi RC} \right)^{-2} \tag{4.42}$$

where τ_{tr} represents the carrier transit time, and R represents the total series resistance (e.g., series resistance of detectors, load resistance, and contact resistance). According to the work of Zhuang and his cooperators, the 3 dB cut-off frequency under X-ray illumination relates to device's parallel or perpendicular structure. The device with perpendicular direction shows a faster response because its thickness is smaller than that of the parallel. Therefore, they pointed out that the perpendicular architecture has more potential in obtaining a detector with low detection limit and acceptable sensitivity.

Contrast: Contrast is a basic metric in imaging to distinguish close structures with similar attenuation lengths. It can be calculated by comparing the modulation of the brightest and darkest part in the sinusoidal grating. The formula is as follows:

$$CR = \frac{(I_b - I_0)}{(I_b + I_0)} \tag{4.43}$$

where I_b and I_0 represent the different intensities of the background and object, respectively. For imaging, lower CR value leads to a smoother edge with less detected details, while high CR value shows sharper edges with more details. The contrast of an image is usually limited by the generation of isotropic light, the absorber layer, and optical crosstalk. The optical crosstalk could be light deflection and scattering within adjacent pixels, which would reduce color and luminance resolution. Additionally, the size of the pixel also limits the contrast. Typically, the pixel size should be smaller than 10 μm for the application in commercial imaging. For small object size, a larger contrast is required.

Modulation transfer function (MTF): MTF is a key parameter for X-ray imager, and typically used to describe the spatial resolution of the X-ray detector, which represents the transmission ability of the input signal modulation of the spatial frequency (unit: line pairs per millimeter (lp mm^{-1})). When MTF is 0.2, the larger the spatial resolution is, the clearer the image of the detector. Typically, the spatial resolution should be larger than 10 lp mm^{-1} for medical imaging [77]. The formula of MTF can be written as follows [78]:

$$\text{MTF}(f) = T_w(f) \times T_{tr}(f) \times T_a(f) \tag{4.44}$$

$$T_w(f) = P_{Pe}T_{pe} + P_k T_k \tag{4.45}$$

where T_{pe} and T_k represent the primary photoelectron and the k-fluorescence reabsorption. Besides, P_{pe} and P_k represent the probabilities of released carriers from primary photoelectron and the k-fluorescence reabsorption. $T_w(f)$ is the weighted MTF, which stems from the primary photoelectron and the k-fluorescence reabsorption. $T_{tr}(f)$ represents the charge carrier trapping-influenced MTF, and $T_a(f)$ represents the pixel electrodes-induced MTF. Usually, the slanted-edge method is used to measure the MTF in a quantitative resolution value. The MTF decreases as the spatial frequency increases. The spatial frequency is determined when MTF = 0.2 [79].

4.4.3.2 *Device design*

Device designs by tuning the work function or generating different barriers at the interface of metal-semiconductor or p-type semiconductor or n-type semiconductor, which could minimize the leakage current and improve the sensitivity of X-ray detector. As depicted in Figure 4.32(a), there are two different kinds of architectures for X-ray detectors, namely type I symmetric electrode, and type II asymmetric electrode [80]. For the asymmetric electrode architecture, a higher Schottky barrier would appear at the interface of Ga and MAPbI$_3$ SC under reverse bias, which significantly suppresses the leakage current because of insufficient carrier injection. Differently, in symmetric electrode architecture, small Schottky barrier for holes is formed at the semiconductor–metal junction, which makes holes inject into the MAPbI$_3$ SC easily, and causes large leakage current.

For perovskite SCs, a larger Schottky barrier would form in the electrode/perovskite interface because Au ($\Phi_{Au} = 5.1$ eV) possesses a higher work function than Al ($\Phi_{Al} = 4.28$ eV). The large barrier tends to suppress ion migration and further decreases leakage current under the bias. Based on this idea, some meaningful work has appeared to solve the uncontrollable ions migration problem. Xu *et al.* sputtered a layer of Al on the top of MAPbBr$_3$ SC to construct a Schottky barrier in the Al/MAPbBr$_3$ interface. After comparing with the Au/MAPbBr$_3$/Au structure device, they found that the Schottky barrier could greatly suppress leakage current at high electric fields (1.47×10^7 V m^{-1}), which optimized the sensitivity from 62 μC Gy$_{air}^{-1}$ cm^{-2} to 359 μC Gy$_{air}^{-1}$ cm^{-2}. For perovskite films, Kim *et al.* introduced an engineering technique to modify the work function by tactfully employing polyimide (PI)-MAPbBr$_3$ composites and PI-MAPbI$_3$ composites as hole-blocking layer (HBL) and hole-transporting layer (HTL), respectively. Because the large intrinsic resistivity of PI reduced the possibility of electrical crosstalk among the pixels, the corresponding X-ray detector possessed a competitive ls of 1.0×10^4 cm^2 V^{-1}, and an impressive sensitivity of 11 μC Gy$_{air}^{-1}$ cm^{-2} [81].

4.4.4 *Indirect X-ray Detection*

Scintillators based on indirect X-ray detectors have the advantages of fast response time and convenient integration with TFT or CMOS array, and are the mainstream products in the market. Recent studies suggest that MHPs also have good potential as efficient scintillators. MHPs have tunable bandgaps by adjusting the B/X sites, which can match the photo-detector response wavelength. Previous research showed that the PLQY of halide perovskite nanocrystal reached nearly 100%, along with the fast-luminescent decay process (~nanoseconds), which enables low afterglow for X-ray imaging.

Afterglow: The afterglow is the radioluminescence intensity at a given time after X-ray radiation. The afterglow has a different requirement with different applications and CT has the highest requirement of 0.1% @ 3 ms [82]. Long afterglow can lead to imaging artifacts because of the signal superimposition on the previous exposure. As shown in Figure 4.18(a), Heo *et al.* compared the afterglow and response time of $CsPbBr_3$ nanocrystal and GOS [79]. It turns out that the afterglow of $CsPbBr_3$ nanocrystal was about 1% @ 30 ns, much better than conventional scintillator GOS. Besides, some low-dimensional perovskites, such as $(n\text{-}C_6H_{13}NH_3)_2PbI_4$, possess ultra-fast decay for hundreds of ps. With negligible afterglow, halide perovskites thus show great potential in dynamic X-ray imaging. Zhu *et al.* reported $Cs_2Ag_{0.6}Na_{0.4}In_{0.85}Bi_{0.15}Cl_6$ scintillator for dynamic X-ray imaging [82]. The afterglow of $Cs_2Ag_{0.6}Na_{0.4}In_{0.85}Bi_{0.15}Cl_6$ is 0.1% @ 16 ms and much lower than the widely used scintillator CsI:Tl (1.5% @ 3 ms).

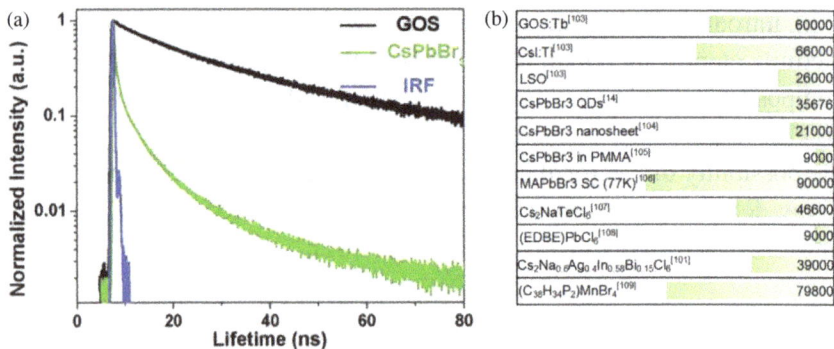

Figure 4.18.   (a) Comparison of the luminance decay time of $CsPbBr_3$ nanocrystals and GOS. The luminance intensity of $CsPbBr_3$ nanocrystals is about 1% after 30 ns. (b) Summary of LY for MHPs and other scintillators commonly used in FPD. Reproduced from Ref. [73].

Light yield: Light yield (LY) is another crucial parameter for scintillators, which decides the sensitivity and detection limit of the detector. The maximum LY of a scintillator can be evaluated by the above equation: $LY = 10^6/\Delta$, with the unit of photons/MeV. Compared with other scintillators in X-ray detection, halide perovskites have high theoretical LY of approximately 2×10^5 photons/MeV due to their defect tolerance and relatively small band gap. However, the actual LY will be reduced by self-absorption, non-radiative recombination, and non-unity light extraction efficiency [77]. In Figure 4.18(b), some LY of halide perovskites together with conventional scintillators are summarized. Due to the self-absorption effect, the created photons could be reabsorbed by the scintillator itself and thus decrease the LY. Furthermore, defects in the scintillator cause non-radiative recombination, which further decrease the LY. In a practical application, a reflective coating is used to avoid light loss, and some optical coupling glue may be applied between the scintillator and photodiode to guarantee the light extraction efficiency.

Emission wavelength: The emission wavelength of scintillator also needs to be optimized to maximize LY. α-Si based *p-i-n* photodiode generally reaches the highest responsivity within the wavelengths of 500–600 nm. It is desired if the emission wavelength of the scintillator matches the responsivity peak of the photodiode. For such a purpose, the emission wavelength of halide perovskites can be adjusted continually by tuning the B/X sites. Chen *et al.* adjusted the X site of $CsPbX_3$ (X = Cl, Br, I) nanocrystals and fabricated different scintillators with emission covering from UV to red [83].

Scalable fabrication methods

Unlike visible light, it is difficult to reflect, diffract, or concentrate X-rays, and panels with large areas are hence required for X-ray imaging. Halide perovskites can be easily fabricated through low-cost methods, such as nanocrystal film [83], glass–ceramic scintillator [84], ceramic scintillator [85], and flexible scintillator screen [86], as shown in Figure 4.19.

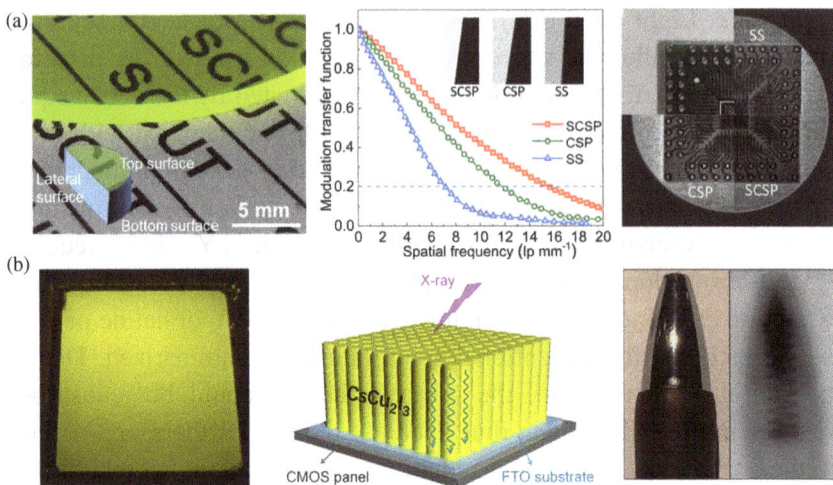

Figure 4.19. (a) Diagonal view of the polished TPP_2MnBr_4 ceramic scintillator with thicknesses of 1 mm under ambient light. Modulation transfer functions (MTF) of X-ray images and photographs of a circuit board. Reproduced from Ref. [85]. (b) Photoluminescent photograph of oriented structured $CsCu_2I_3$ thick film. Schematic representation of $CsCu_2I_3$ scintillator detector. Physical picture and X-ray image of a ballpoint pen based on the $CsCu_2I_3$ scintillator detector. Reproduced from Ref. [87].

4.4.5 *Challenges and Outlook*

While halide perovskites have shown promise for X-ray detectors, there are still several challenges to be tackled before they can be used for commercial applications, such as low quality and stability of perovskites, small area, insufficient thickness, and uncontrollable ionic migration. Therefore, there is urgent need for further research [88].

4.4.5.1 *Challenges*

Improving stability of perovskite materials: It is known to all that though perovskite materials have achieved extraordinary progress in PSCs, and are reconsidered as the next-generation energy, their inherent instability is a great barrier for not only PSCs but also X-ray detectors. For organic–inorganic hybrid perovskites, the organic A-site cations (FA^+, MA^+ and their mixtures) are volatile and hygroscopic, which causes perovskites to decompose under extreme environments (heating, humidity, and UV-light, etc.) [89]. For inorganic perovskites, though they showed a better thermodynamic stability than hybrid ones, they still suffer from a phase transition problem. Therefore, stability problem should be conquered for further application in photoelectric devices.

Suppressing ionic migration: Ionic migration is a notorious problem in halide perovskite, because both the output signal and noise are directly related to it [90]. The uncontrollable ionic migrations not only induce baseline drifting, influence the signal recording, and trigger overflow error, they also induce ionic conductivity, increase dark current, and enlarge the shot noise. All of these damage the sensitivity, LoD, and the stability of detectors, because of which ionic migration should be effectively suppressed.

Improving the resolution of perovskite based-imagers: For recent perovskite-based X-ray imagers, the spatial resolution of advanced $MA_3Bi_2I_9$ SC is 4.22 lp mm^{-1}, $Cs_2AgBiBr_6$ wafer is 4.9 lp mm^{-1}, and $MAPbI_3$ film is 3.1 lp mm^{-1}. All of them are smaller than the scintillator ($CsPbBr_3$ nanocrystals)-based indirect X-ray detector (9.8 lp mm^{-1}), and the commercial applied GOS-based detectors with spatial resolution = 6.2 lp mm^{-1} at MTF = 0.2. This cannot meet some of the requirements of commercial application of medical imaging, whose resolution requirement is beyond 10 lp mm^{-1}.

Growth of high-quality perovskite with reproducibility: The quality of perovskite fluctuates greatly. The resistivity of the perovskite is in a wide range from 10^7 cm to 10^{12} cm, which is unsuitable to fabricate high-sensitivity X-ray detectors with reproducibility. Notably, high sensitivity is especially important in reducing health risk by minimizing X-ray exposure time during routine medical inspections, which is also related to the quality of materials.

Fabricating perovskite in large areas, and to be flexible with comparable thickness: Generally, the X-ray refractive index is about 1.0 and difficult to focus, so large-area X-ray detectors with the size comparable to the object are necessary for imaging applications. Besides, flexibility is significant property for non-planar surfaces' detection. However, this is more challenging in SC perovskites, because of limited technology and expensive cost. Though large area and flexibility can be achieved in perovskite films, the quality and thickness remain great problems. The thickness of perovskite layer should be large enough to efficiently absorb X-ray emissions. Typically, the thickness of $MAPbBr_3$ to completely absorb X-ray with 22 keV should reach 2 mm, which is difficult to achieve in film deposition technology [91]. Therefore, it's challenging to guarantee both thickness and large area in perovskite films.

4.4.5.2 *Outlook*

Fabricating low-dimension perovskite materials: Although the 2D perovskite has low density (resulting in lower stopping power and carrier properties), it still is a potential direction for commercial X-ray detectors. One reason is that reducing dimensions tends to increase the activity energy against the uncontrollable carrier's migration in halide perovskite. Additionally, a low-dimensional perovskite possesses a more symmetric crystal structure with small surface energy. Both of them can improve the performance and lifetime of the device. Meanwhile, 2D perovskite would be introduced with large organic molecules to cut down its 3D structure, which increases the bulk resistance to avoid noise and further improves the sensitivity of the detector. QDs/NCs can be used as ink for fabricating perovskite films with sufficient thickness and large area. However, because the organic ligand was introduced, the transport issue of QDs/NCs should be conquered [92].

Introducing passivated layers or interfacial design to minimize the influence of ion migration: According to early reports, ion migration mostly occurs along the GBs through the defective sites. Thereby, passivating the grain boundary or interfacial design to decrease the defect state is a useful way to increase the activation energy of ion migration for reducing ionic migrations.

Developing inorganic perovskite-based X-ray detector: Using inorganic perovskite, with endurance Cs^+ ion replacing the organic ion of hybrid perovskites, can improve the material's stability under ionizing radiation. Besides, the Cs^+ ion formed perovskite ($CsPbI_3$, $CsPbBr_3$ and $Cs_2AgBiBr_6$)

shows larger Z value than MA^+ or FA^+ formed perovskite ($MAPbI_3$, $MAPbBr_3$), which indicates better ability of absorbing X-rays of a certain thickness. Moreover, the resistivity of inorganic perovskites is typically larger than that of organic ones. The resistivity of inorganic SCs ($CsPbI_3$ SC, $CsPbBr_3$ SC, and $Cs_2AgBiBr_6$ SC) and organic–inorganic hybrid SCs ($MAPbI_3$ SC and $MAPbBr_3$ SC) are in ranges of 10^8–10^{12} cm and 10^7–10^8 cm, respectively. Therefore, inorganic perovskites could be a promising direction of research.

Structural design and fabricated method optimization: The biggest obstacle in obtaining high sensitivity and small LoD is the dark current, which inevitably induces noise and deteriorates X-ray detector performance. As we mentioned above, the LoD is defined when SNR is 3. Hence, there are two main directions: first, increasing the intensity of the signal, which can be achieved by fabricating high-quality perovskite SC, film or wafer, with continually optimized technologies; second, decreasing the dark current and noise directly. As we mentioned in the section on "Device design", designing devices structurally by building different barriers is a useful strategy.

Developing large atomic number perovskite and optimizing film deposition technologies: It is difficult to combine large area and thickness in the preparation of perovskite films in spin-coating method. Typically, the thickness of perovskite film cannot reach the required value (several-hundred micrometer thickness) to completely absorb X-ray emissions. One promising solution is to increase the atomic number, because the large atomic number can stop the X-ray with large stopping power. Additionally, developing new and improved film deposition technologies (e.g., inkjet printing, CH_3NH gas assisted, D-bar coating, and slot-die coating) is also an important future direction.

4.5 Fluorescent Sensors and Thermometers

4.5.1 *Fluorescent Sensors*

The key parameters, including the response time, recyclability, and detection range, are considered as the performance evaluation standard for fluorescent sensors. The unique emission characters of MHPs are promising for applications in this field [93,94]. As a typical case, Su *et al.* reported that the switchable dual emission between the hydrated and dehydrated crystal endows $Cs_2InBr_5 \cdot H_2O$ with the intriguing capability to act as a model PL water-sensor [95]. When the PL sensor is exposed to humid air that has 80% relativity humidity (RH), the emission wavelength and intensity are gradually varied, and the corresponding response time (the color changes from the original yellow to red) is 30 s. In the case of low RH of about 30%, the response time is prolonged to 120 s, indicating an accelerated response speed in high RH. More importantly, $Cs_2InBr_5 \cdot H_2O$ shows a superior response speed than that of other all-inorganic materials. Based on that, the recyclability tests of the sensing material correlating to the water release/uptake are further performed in ambient atmosphere, as shown in Figure 4.35(c). An excellent PL representative prototype, through direct suction of dehydrated powder into THF solvent with different water content, are carried out via subsequent water detection test, as shown in Figure 4.35(d). Accordingly, the *in situ* transformation between hydrated $Cs_2InBr_5 \cdot H_2O$ and the dehydrated form is accompanied with a switchable dual emission, which favorably enables it as a PL water-sensor in humidity detection or traces of water detection in organic solvents.

Recently, Xia's group designed and synthesized a novel 0D lead-free metal halide of $(C_9NH_{20})_2MnBr_4$ with a high PLQY of around 81.08% and a green emission peaking at 528 nm, which is originated from the $^4T_1$ to $^6A_1$ radiative transition of Mn^{2+} ions in $[MnBr_4]$ unit [94]. Moreover, $(C_9NH_{20})_2MnBr_4$ possesses an excellent moisture, heat, and light stability. As we all know, the sudden quenching or shining of luminescence can be used as an effective way to detect toxic and volatile gases. Interestingly, the emission of $(C_9NH_{20})_2MnBr_4$ can be quickly quenched by acetone vapor, and thus we explored their potential applications in this field of fluorescent sensor. The gas-sensitive detecting device diagram is shown in Figure 4.20(a). By detecting the PL emission before exposure to acetone vapor and 10 s after exposure to acetone vapor (Figure 4.20(b)), the strong green emission is reduced by more than 50 times within 10 s. This good switchability is the key to gas-sensitive applications, and the selectivity is another important indicator. The fluorescent responses of $(C_9NH_{20})_2MnBr_4$ upon exposure to different organic solvent vapors for 10 min are measured,

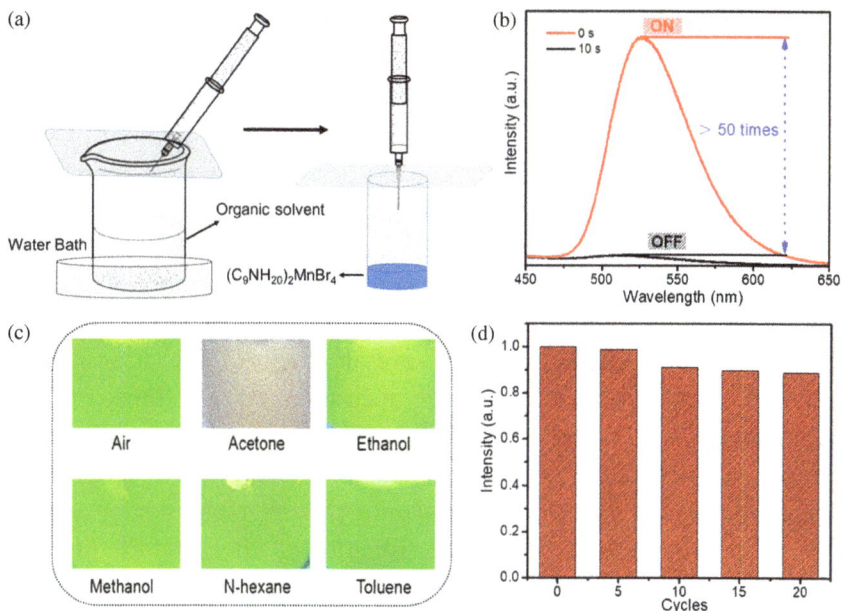

Figure 4.20. (a) Schematic diagram of a simplified device for gas sensing application. (b) Time-related emission spectra of $(C_9NH_{20})_2MnBr_4$ toward acetone vapor. (c) Digital photos of $(C_9NH_{20})_2MnBr_4$ upon exposure to various organic vapors for 10 min under 365 nm UV excitation. (d) PL intensity of $(C_9NH_{20})_2MnBr_4$ toward acetone vapor for 20 cycles. Reproduced from Ref. [94].

and the corresponding digital photos under 365 nm excitation are presented in Figure 4.20(c). Except for acetone vapor, no obvious fluorescence variation is observed, indicating that $(C_9NH_{20})_2MnBr_4$ has selective fluorescence quenching effect on acetone vapor. Importantly, the PXRD patterns show no signs of material decomposition after 20 cycles, and the emission intensity does not decrease significantly (Figure 4.20(d)), which indicates that $(C_9NH_{20})_2MnBr_4$ has good reversibility and circulation as a potential candidate for the fluorescent sensor application.

4.5.2 *Thermometers*

Thermometer endows numerous applications ranging from medicine and defense, to biological research itself or the diagnosis of technical failures.

Although MHPs have revolutionized research in optoelectronics owing to the unique optical performances, their lower-dimensional counterparts can further expand the fields with unprecedented functionalities. Kovalenko *et al.* presented the strong temperature dependence of lifetimes in low-dimensional MHPs, and applied these characteristics to thermography. In particular, Cs_4SnBr_6 shows a green broad-band emission centered at 535 nm at RT, with a FWHM of 120 nm, and a PLQY of 20% [96]. There is a strong correlation between lifetime and temperature, and the lifetimes have highly reproducible variation, which makes similar MHPs with high thermometric precision. Although several time-resolved measurement techniques could be used to precisely measure the PL lifetime, all have traditionally been limited by the fact that they only use a single channel detector. Consequently, lengthy acquisition through point-by-point scanning is required, which severely limits the thermographic image capture rate. As a solution to this problem, ToF fluorescence lifetime imaging (FLI) technique was used to acquire a 2D map of PL lifetimes, and combined with thermally sensitive luminophores. The working principle for such a ToF-FLI image sensor is based on the acquisition of four phase-locked images at 0°, 90°, 180°, and 270° phase differences with respect to the excitation signal, I_0, I_1, I_2, and I_3 followed by recalculation of the average intensity I, modulation depth M, and phase delay $\Delta\Phi$. For the test of the thermographic performance, a Cs_4SnBr_6 powder was deposited over a resistively heated pattern and enclosed the material with a glass coverslip, and then measured the resulting lifetime image of the heated pattern. Compared to a conventional bolometric thermogram, the ToF-FLI method showed a much higher lateral thermographic resolution.

4.6 Photocatalysis

Solar-driven water splitting provides a leading approach to store the abundant yet intermittent solar energy and produce hydrogen as a clean and sustainable energy carrier. A straightforward route to light-driven water splitting is to apply self-supported particulate photocatalysts, which is expected to be competitive with fossil-fuel-derived hydrogen generation. More importantly, the powder-based systems can lend themselves to making functional panels on a large scale while retaining the intrinsic activity of the photocatalysts. MHPs have emerged as new frontier materials for optoelectronic as well as energy applications. In addition to various well-known applications, such as solar cells, LEDs, photodetectors, and resistive switching memories, MHPs can be utilized as efficient photocatalysts for numerous electrochemical reactions, including carbon dioxide (CO_2) reduction reactions, hydrogen evolution reaction, photosynthesis, and wastewater treatment. However, the use of MHPs toward photocatalysis remains a tremendous challenge owing to their poor stability in polar solvents. Nevertheless, huge progress has been made to counter this critical issue for improving the performance of MHPs as efficient photocatalysts in a wide range of applications.

4.6.1 *Progress of Photocatalysis*

Like the natural and efficient energy conversion exhibited by photosynthesis, photocatalysis presents an effective way for man-made solar-to-chemical energy conversion. The first report of UV-driven photocatalytic hydrogen production on TiO_2 in 1972 has since motivated decades of scientific exploration and development, leading to several applications important for human society; from energy conversion (e.g., water splitting, CO_2 reduction) and chemical transformations, to the remediation of organic pollutants. Reactions driven by photocatalytic materials generally

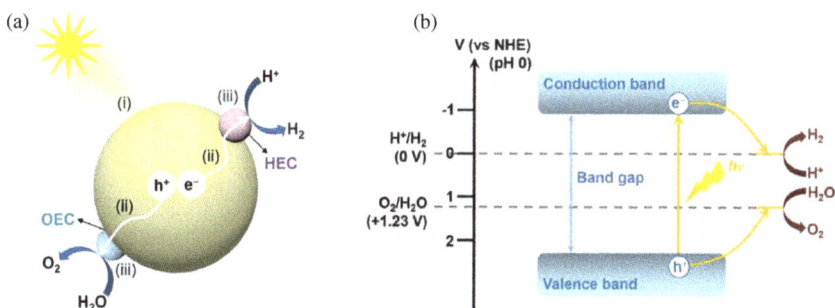

Figure 4.21. (a) Diagram showing the reactions during water splitting on a semiconductor photocatalyst: (i) light absorption, (ii) charge separation and transport, and (iii) redox reactions. (b) Energy diagram for photocatalytic water splitting based on one-step excitation. Reproduced from Ref. [97].

follow three separate processes [97]: (i) absorption of photons to generate electron and hole pairs, (ii) charge separation and migration to reaction sites on the photocatalyst surface, and (iii) chemical oxidation and reduction at the surface mediated by the photogenerated holes and electrons, respectively.

4.6.2 *Advantages of Metal Halide Perovskites for Photocatalysis*

An ideal photocatalytic material should embody several desirable physical traits, namely, broad and strong light absorption, efficient charge separation, long operational stability, and the appropriate redox ability for target reactions. In reality, however, most photocatalysts are far from perfect. For example, pristine TiO_2, graphitic carbon nitride (g-C_3N_4), and $BiVO_4$ have their own shortcomings such as wide bandgap, rapid recombination of photogenerated charge carriers, and poor photoreduction potential, respectively. It follows that a strong desire to develop new semiconductor photocatalysts with suitable properties continues to motivate intense materials exploration and research within the field. MHPs exhibit relatively narrow bandgap energies as shown in Figure 4.22, enabling the absorption of lower-energy solar photons.

Based on the reaction thermodynamics, a suitable match is required between the electronic band structure of a semiconductor and reaction

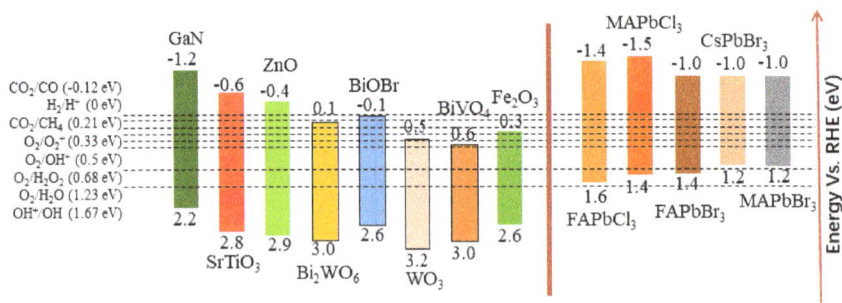

Figure 4.22. Band edge positions of conventional photocatalysts and different MHPs relative to reversible hydrogen electrode. For comparison, the redox potential of some common half-reactions is also presented.

redox potential. The potentials of typical photocatalytic half-reaction involved in water splitting, CO_2 reduction, aerobic organic transformation, and dye degradation are shown in Figure 4.22 [98]. Here, the relative position of the MHP CB and VB are depicted alongside these redox potentials. Based on their comparison, Figure 4.22 highlights the excellent reduction ability of MHPs; that is, the relative position of their CB is typically negative enough for H_2 generation, CO_2 reduction, superoxide radical generation. Additionally, some members of the MHP family (namely the Cl-based materials and all-inorganic $CsPbBr_3$) can in theory also achieve water oxidation, because of their relatively positive valence band maximum (VBM). For hydroxyl radical formation, which is often involved in dye degradation, the required potential is above the VBM of MHPs, making them an unsuitable candidate.

4.6.3 *MHP Photocatalysis*

4.6.3.1 *CO_2 reduction reaction*

Photocatalysis not only reduces the currently increasing CO_2 emissions but also could convert CO_2 into chemical fuels, such as methane and methanol. When MHPs were applied for CO_2 photo-reduction as catalysts, their position of CB should be above the redox potential of the CO_2. The major gaseous products such as CH_4 and CO can be produced on the surface of the catalyst as in the following equations

$$CO_2 + 2H^+ + 2e^- \rightarrow CO + H_2O, E^\circ \text{ redox} = -0.51 \text{ V} \qquad (4.46)$$

$$CO_2 + 8H^+ + 8e^- \rightarrow CH_4 + 2H_2O, E^\circ \text{ redox} = -0.24 \text{ V} \qquad (4.47)$$

Therefore, with suitable band structures of MHPs, the excited electrons at CB can facilitate redox reaction before undergoing recombination. The catalytic activity of inorganic HP QDs, in particular, $CsPbBr_3$, was reported in 2017 by Hou *et al.* who investigated the dependence of the CO_2 reduction efficiency on the $CsPbBr_3$ QDs size. Due to their activation in a large portion of the solar spectrum and suitable bandgap possession, $CsPbBr_3$ is a promising photocatalytic semiconductor. With the increase of QDs' size, the bandgap and PL spectra were tunable over the visible spectral region by

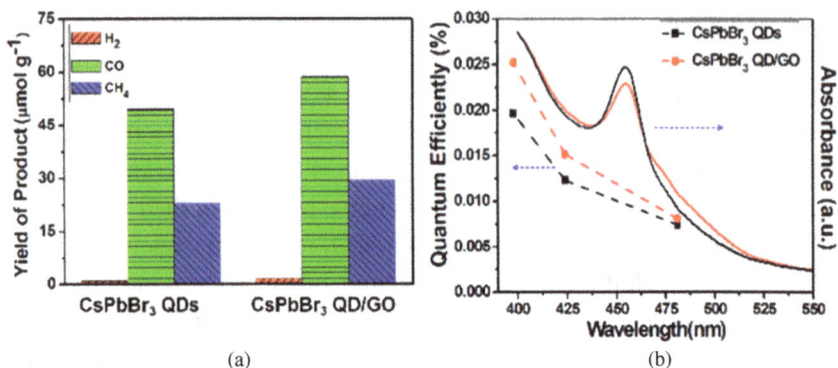

Figure 4.23. (a) Photocatalytic performance: yield of CO_2 reduction reaction products after 12 h of photochemical reaction. (b) UV-visible absorption spectra and external quantum efficiency plots. Reproduced from Ref. [100].

quantum size effects. The increased carrier lifetime, revealed by time-resolved PL spectra, accounts for the improved electron–hole separation efficiency. As a result, the $CsPbBr_3$ QDs with an optimum size achieved high photochemical conversion efficiency for the solar CO_2 reduction reaction. The size of the $CsPbBr_3$ QDs was approximately 4–12 nm depending on synthesis temperature offering high surface area and short charge-transfer pathway. When the particle size increased, the optical absorption edges turned toward longer wavelengths that expanded absorbance. The experimental results indicated that 8.5 nm was the optimal QDs size that achieved the highest yields of 4.3, 1.5, and 0.1 μmol g^{-1} h^{-1} for CO, CH_4, and H_2, respectively. The CBM of $CsPbBr_3$ were — 1.03 V versus the normal hydrogen electrode (NHE) at pH 0, which was more negative than that of $E°(CO_2/CO) = -0.51$ V versus NHE and $E°(CO_2/CH_4) = -0.24$ V versus NHE, indicating that the photogenerated electrons can react with adsorbed CO_2 to produce CO and CH_4 (Figure 4.23) [99,100].

4.6.3.2 *Photodegradation of dyes and contaminants*

Many strategies have been developed for wastewater treatment, including sedimentation and filtration. However, most of them only separate contaminants from liquids, and further treatment is required to convert these contaminants into non-toxic substances.

The photoactivity of MHPs for organic dyes degradation was first reported by Gao *et al.* in 2017. All-inorganic $CsPbX_3$ MHPs were prepared using the emulsion and de-emulsion method at room temperature, and then various amounts of $CsPbX_3$ were utilized for the photocatalytic degradation of methyl orange (MO), a common organic pollutant, from solutions. When 0.1 mg $CsPbCl_3$ was used, the percentage of MO degradation was 50%, which increased to 90% within 80 min when 2 mg $CsPbCl_3$ was used as a photocatalyst. While under visible light irradiation without $CsPbX_3$, only 17% of MO was degraded within 80 min, which proved the photocatalytic degradation activity of $CsPbX_3$. When $CsPbX_3$ was irradiated by the light, the photo-excited electrons and holes were created in CB and VB separated by the bandgap. The redox potential of H_2O (OH^-)/$^{\cdot}OH$ and O_2/$^{\cdot}O_2^-$ was within the bandgap of $CsPbX_3$ so O_2 can receive electrons to form $^{\cdot}O_2^-$ and H_2O can be transferred to $^{\cdot}OH$ by receiving holes. These $^{\cdot}O_2^-$ and $^{\cdot}OH$ radicals were responsible for dye degradation.

Zhang *et al.* improved the stability of HPs and successfully developed an alcohol-based photocatalytic system based on $Cs_2AgBiBr_6$, a lead-free HP [101]. Several ionic dyes, including Rhodamine B (RhB), Rhodamine 110 (Rh110), methyl red (MR), and MO, were used to evaluate the photocatalytic activity of $Cs_2AgBiBr_6$. RhB was almost completely degraded (up to 98%) within 120 min of irradiation while for the other dyes, the degradation occurred much slower. The efficiency of the $Cs_2AgBiBr_6$ photocatalyst increased when noble metals, such as Pt and Au, were deposited on its surface to facilitate the photogenerated carrier separation, which was evidenced by the PL quenching.

4.6.3.3 *Photosynthesis of organic compounds*

Despite their wide applications as photocatalysts for CO_2RR or dye and contaminant degradation, lead HPs have been rarely utilized for the photosynthesis of organic compounds. However, some papers on lead HPs as photosynthesis catalysts have recently been published. Zhu *et al.* demonstrated that $APbBr_3$ (A = Cs or MA) HP colloids could be effective photocatalysts for C–C bond formation reactions, by exploring a model reaction: the α-alkylation of aldehydes. Under blue LED irradiation,

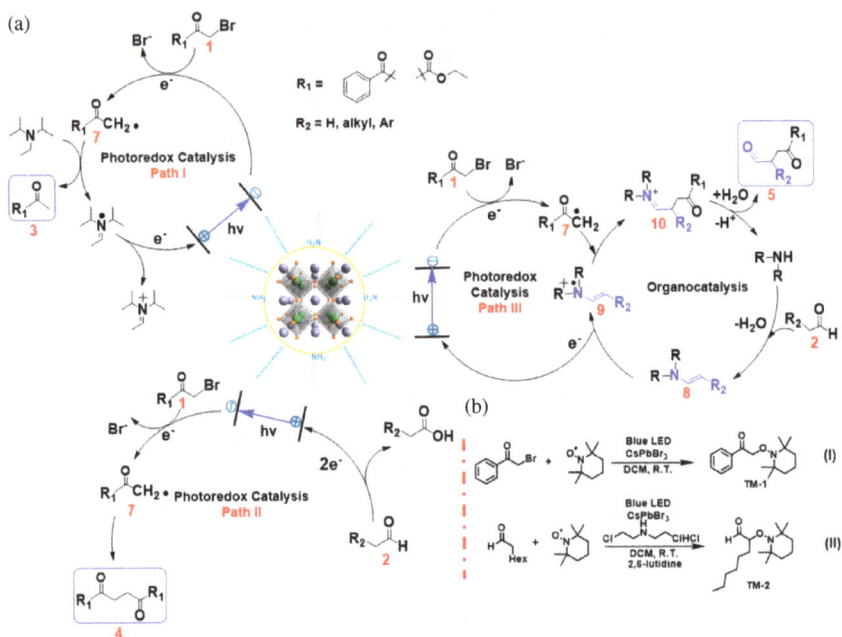

Figure 4.24. (a) Proposed mechanism for perovskite catalyzed dehalogenation, sp^3 C-coupling, and α-alkylation. (b) TEMPO-trapped experiment for radical intermediate validation. Reproduced from Ref. [102].

several products were generated from 2-bromoacetophenone and octanal, including dehalogenated acetophenone (76% yield), sp^3 C-coupling product (8% yield), and α-alkylation product (7% yield) (Figure 4.24) [102]. Furthermore, they indicated that HP crystals could selectively photocatalyze the reaction toward the desired product. The mechanism for these transformations was proposed to be the single electron reduction of 2-bromoacetophenone to form a radical (7 in Figure 4.24(b)), followed by the extraction of an H atom from the co-catalyst or self-couples to generate the products labeled as **3** or **4** in Figure 4.24(b). The regeneration of the HPs catalyst was proposed to occur via the oxidation of the co-catalyst or an aldehyde for paths I and II, respectively.

4.6.3.4 *Hydrogen evolution reaction*

Splitting hydrogen iodide into H_2 and I_3^- in the presence of $MAPbI_3$ under solar irradiation was the first milestone for the photocatalytic application of HPs, which was reported by Park *et al.* In their work, Park and his group exploited the concept of dynamic equilibrium in aqueous HI solution to maintain $MAPbI_3$ stability. $MAPbI_3$ was considered as an ionic crystal consisting of MA^+ and PbI_3^- that could be formed in saturated solutions. Therefore, when $MAPbI_3$ precipitates dissolved in the saturated solution, they could be dissociated into MA^+ and PbI_3^- ions and at the same time MA^+ and PbI_3^- ions were reprecipitated in the crystalline form at an equal rate resulting in $MAPbI_3$ powder that could remain stable in aqueous HI solution.

Wang *et al.* reported a new strategy for improving the photocatalytic H_2 evolution ability of organic–inorganic $MAPbBr_3$ HPs, which involved hybridizing them with Pt/Ta_2O_5 and poly(3,-4-ethylenedioxythiophene): polystyrene sulfonate (PEDOT:PSS) nanoparticles as electron- and hole-transporting channels, respectively [103]. The new system acted as a potential catalyst, which increased the H_2 formation rate approximately 52 times. This was attributed to the increase in charge separation and transportation efficiency of $MAPbBr_3$ by providing new dual charge transporting pathways. Tantalum pentoxide (Ta_2O_5) was considered to be the

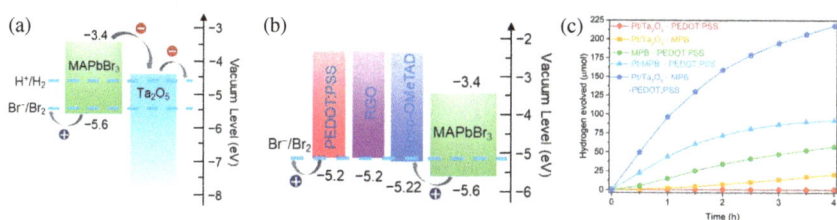

Figure 4.25. (a) Energy level diagrams of methylammonium lead bromide ($MAPbBr_3$) and Ta_2O_5 and redox potentials for HBr splitting reaction. (b) Schematic energy-level diagram of $MAPbBr_3$ and hole-transporting materials (PEDOT:PPS, poly(3,4-ethylenedioxy-thiophene) polystyrenesulfonate; rGO, reduced graphene oxide; Spiro-OMeTAD, 2,20,7,70-tetrakis[N,N-di(4-methoxyphenyl)amino]-9,90-spirobifluorene). (c) Time evolution of photocatalytic H_2 generation on Pt/Ta_2O_5-PEDOT:PSS, Pt/Ta_2O_5-PEDOT:PSS, Pt/Ta_2O_5-MAPbBr_3$, $MAPbBr_3$-PEDOT:PSS, and Pt/Ta_2O_5-MAPbBr_3$-PEDOT:PSS (MPB is $MAPbBr_3$). Reproduced from Ref. [103].

ideal conduction band edge for transferring electrons from $MAPbBr_3$, and thus, Pt/Ta_2O_5-$MAPbBr_3$ contributed to the increase in the rate of the HER (Figure 4.25(a)). Because its VBM was more positive than that of $MAPbBr_3$, PEDOT:PSS was selected as efficient hole-transporting material for the Pt/Ta_2O_5-$MAPbBr_3$ hybrid system, which favored the Br-oxidation reaction (Figure 4.25(b)), thus Pt/Ta_2O_5-$MAPbBr_3$-PEDOT: PSS was the most efficient photocatalytic system for HER (Figure 4.25(c)). The authors also demonstrated that PEDOT:PSS significantly improved the photocatalysis efficiency of the Pt/Ta_2O_5-$MAPbBr_3$ system compared with that of other hole transporting materials, namely rGO, Spiro-OMeTAD, or IrO_2 because it could oxidize Br^- with negligible overvoltage, as confirmed by electrochemical measurements.

4.7 Summary

MHPs have affirmed their pedigree as emerging semiconducting materials, exhibiting unique electronic, optical, and magnetic properties. They have demonstrated tremendous versatilities in terms of structure and composition and potential for emerging technological applications ranging from PV to LEDs, photodetectors, laser, X-ray scintillators, thermometers, and photo-catalysis. Major progress has been made in improving efficiency and stability with compositional engineering. Intrinsic structural stability and long-term environmental stability have improved substantially. Recent developments have focused on compositional engineering, exploring all-inorganic perovskites and Pb-free perovskites, enhancing stability, and scaling up for device applications.

References

[1] W. Zhang, G.E. Eperon, and H.J. Snaith, *Nat. Energy* 1, 16048 (2016).
[2] J.Y. Kim, J.-W. Lee, H.S. Jung, H. Shin, and N.-G. Park, *Chem. Rev.* 120, 7867–7918 (2020).
[3] I. Mesquita, L. Andrade, and A. Mendes, *Renew. Sust. Energ. Rev.* 82, 2471–2489 (2018).
[4] A.K. Jena, A. Kulkarni, and T. Miyasaka, *Chem. Rev.* 119, 3036–3103 (2019).

[5] Z. Li, M. Yang, J.-S. Park, S.-H. Wei, J.J. Berry, and K. Zhu, *Chem. Mater.* 28, 284–292 (2016).

[6] W.-J. Yin, T. Shi, and Y. Yan, *J. Phys. Chem. C* 119, 5253–5264 (2015).

[7] J.-W. Lee, D.-K. Lee, D.-N. Jeong, and N.-G. Park, *Adv. Funct. Mater.* 29, 1807047 (2019).

[8] P. Zhao, B.J. Kim, X. Ren, D.G. Lee, G.J. Bang, J.B. Jeon, W.B. Kim, and H.S. Jung, *Adv. Mater.* 30, 1802763 (2018).

[9] G.S. Han, J. Kim, S. Bae, S. Han, Y.J. Kim, O.Y. Gong, P. Lee, M.J. Ko, and H.S. Jung, *ACS Energy Lett.* 4, 1845–1851 (2019).

[10] D. Bi, W. Tress, M.I. Dar, P. Gao, J. Luo, C. Renevier, K. Schenk, A. Abate, F. Giordano, J.-P. Correa Baena, J.-D. Decoppet, M. Zakeeruddin Shaik, K. Nazeeruddin Mohammad, M. Grätzel, and A. Hagfeldt, *Sci. Adv.* 2, e1501170 (2016).

[11] S. Shin Seong, J. Yeom Eun, S. Yang Woon, S. Hur, G. Kim Min, J. Im, J. Seo, H. Noh Jun, and I. Seok Sang, *Science* 356, 167–171 (2017).

[12] W. Nie, H. Tsai, R. Asadpour, J.-C. Blancon, J. Neukirch Amanda, G. Gupta, J. Crochet Jared, M. Chhowalla, S. Tretiak, A. Alam Muhammad, H.-L. Wang, and D. Mohite Aditya, *Science* 347, 522–525 (2015).

[13] A. Ng, Z. Ren, H. Hu, P.W.K. Fong, Q. Shen, S.H. Cheung, P. Qin, J.-W. Lee, A.B. Djurišić, S.K. So, G. Li, Y. Yang, and C. Surya, *Adv. Mater.* 30, 1804402 (2018).

[14] X. Li, D. Bi, C. Yi, J.-D. Décoppet, J. Luo, M. Zakeeruddin Shaik, A. Hagfeldt, and M. Grätzel, *Science* 353, 58–62 (2016).

[15] B. Ding, Y. Li, S.-Y. Huang, Q.-Q. Chu, C.-X. Li, C.-J. Li, and G.-J. Yang, *J. Mater. Chem. A* 5, 6840–6848 (2017).

[16] A. Ummadisingu and M. Grätzel, *Sci. Adv.* 4, e1701402 (2018).

[17] K. Liang, D.B. Mitzi, and M.T. Prikas, *Chem. Mater.* 10, 403–411 (1998).

[18] N.J. Jeon, J.H. Noh, Y.C. Kim, W.S. Yang, S. Ryu, and S.I. Seok, *Nat. Mater.* 13, 897–903 (2014).

[19] N. Pellet, P. Gao, G. Gregori, T.-Y. Yang, M.K. Nazeeruddin, J. Maier, and M. Grätzel, *Angew. Chem. Int. Ed. Engl.* 53, 3151–3157 (2014).

[20] S. Yang Woon, H. Noh Jun, J. Jeon Nam, C. Kim Young, S. Ryu, J. Seo, and I. Seok Sang, *Science* 348, 1234–1237 (2015).

[21] N. Ahn, D.-Y. Son, I.-H. Jang, S.M. Kang, M. Choi, and N.-G. Park, *J. Am. Chem. Soc.* 137, 8696–8699 (2015).

[22] Y. Zhang, S.-G. Kim, D. Lee, H. Shin, and N.-G. Park, *Energy Environ. Sci.* 12, 308–321 (2019).

[23] Z. Fan, H. Xiao, Y. Wang, Z. Zhao, Z. Lin, H.-C. Cheng, S.-J. Lee, G. Wang, Z. Feng, W.A. Goddard, Y. Huang, and X. Duan, *Joule* 1, 548–562 (2017).

[24] K.A. Bush, K. Frohna, R. Prasanna, R.E. Beal, T. Leijtens, S.A. Swifter, and M.D. McGehee, *ACS Energy Lett.* 3, 428–435 (2018).

[25] S. Gharibzadeh, B. Abdollahi Nejand, M. Jakoby, T. Abzieher, D. Hauschild, S. Moghadamzadeh, J.A. Schwenzer, P. Brenner, R. Schmager, A.A. Haghighirad, L. Weinhardt, U. Lemmer, B.S. Richards, I.A. Howard, and U.W. Paetzold, *Adv. Energy Mater.* 9, 1803699 (2019).

[26] C.O. Ramírez Quiroz, G.D. Spyropoulos, M. Salvador, L.M. Roch, M. Berlinghof, J. Darío Perea, K. Forberich, L.-I. Dion-Bertrand, N.J. Schrenker, A. Classen, N. Gasparini, G. Chistiakova, M. Mews, L. Korte, B. Rech, N. Li, F. Hauke, E. Spiecker, T. Ameri, S. Albrecht, G. Abellán, S. León, T. Unruh, A. Hirsch, A. Aspuru-Guzik, and C.J. Brabec, *Adv. Funct. Mater.* 29, 1901476 (2019).

[27] K.A. Bush, A.F. Palmstrom, Z.J. Yu, M. Boccard, R. Cheacharoen, J.P. Mailoa, D.P. McMeekin, R.L.Z. Hoye, C.D. Bailie, T. Leijtens, I.M. Peters, M.C. Minichetti, N. Rolston, R. Prasanna, S. Sofia, D. Harwood, W. Ma, F. Moghadam, H.J. Snaith, T. Buonassisi, Z.C. Holman, S.F. Bent, and M.D. McGehee, *Nat. Energy* 2, 17009 (2017).

[28] F. Sahli, J. Werner, B.A. Kamino, M. Bräuninger, R. Monnard, B. Paviet-Salomon, L. Barraud, L. Ding, J.J. Diaz Leon, D. Sacchetto, G. Cattaneo, M. Despeisse, M. Boccard, S. Nicolay, Q. Jeangros, B. Niesen, and C. Ballif, *Nat. Mater.* 17, 820–826 (2018).

[29] A. De Almeida, B. Santos, B. Paolo, and M. Quicheron, *Renew. Sust. Energ Rev.* 34, 30–48 (2014).

[30] V. Wood and V. Bulović, *Nano. Rev.* 1, 5202 (2010).

[31] N.Y. Morgan, C.A. Leatherdale, M. Drndić, M.V. Jarosz, M.A. Kastner, and M. Bawendi, *Phys. Rev. B* 66, 075339 (2002).

[32] S. Reineke, F. Lindner, G. Schwartz, N. Seidler, K. Walzer, B. Lüssem, and K. Leo, *Nature* 459, 234–238 (2009).

[33] T. Hattori, T. Taira, M. Era, T. Tsutsui, and S. Saito, *Chem. Phys. Lett.* 254, 103–108 (1996).

[34] J. Byun, H. Cho, C. Wolf, M. Jang, A. Sadhanala, R.H. Friend, H. Yang, and T.-W. Lee, *Adv. Mater.* 28, 7515–7520 (2016).

[35] N. Wang, L. Cheng, R. Ge, S. Zhang, Y. Miao, W. Zou, C. Yi, Y. Sun, Y. Cao, R. Yang, Y. Wei, Q. Guo, Y. Ke, M. Yu, Y. Jin, Y. Liu, Q. Ding, D. Di, L. Yang, G. Xing, H. Tian, C. Jin, F. Gao, R.H. Friend, J. Wang, and W. Huang, *Nat. Photo.* 10, 699–704 (2016).

[36] M. Yuan, L.N. Quan, R. Comin, G. Walters, R. Sabatini, O. Voznyy, S. Hoogland, Y. Zhao, E.M. Beauregard, P. Kanjanaboos, Z. Lu, D.H. Kim, and E.H. Sargent, *Nat. Nanotechnol.* 11, 872–877 (2016).

[37] L. Protesescu, S. Yakunin, M.I. Bodnarchuk, F. Krieg, R. Caputo, C.H. Hendon, R.X. Yang, A. Walsh, and M.V. Kovalenko, *Nano Lett.* 15, 3692–3696 (2015).

[38] S.T. Ochsenbein, F. Krieg, Y. Shynkarenko, G. Rainò, and M.V. Kovalenko, *ACS Appl. Mater. Interfaces* 11, 21655–21660 (2019).

[39] K. Lin, J. Xing, L.N. Quan, F.P.G. de Arquer, X. Gong, J. Lu, L. Xie, W. Zhao, D. Zhang, C. Yan, W. Li, X. Liu, Y. Lu, J. Kirman, E.H. Sargent, Q. Xiong, and Z. Wei, *Nature* 562, 245–248 (2018).

[40] T. Chiba, Y. Hayashi, H. Ebe, K. Hoshi, J. Sato, S. Sato, Y.-J. Pu, S. Ohisa, and J. Kido, *Nat. Photo.* 12, 681–687 (2018).

[41] Y.-H. Kim, G.-H. Lee, Y.-T. Kim, C. Wolf, H.J. Yun, W. Kwon, C.G. Park, and T.W. Lee, *Nano Energy* 38, 51–58 (2017).

[42] G. Niu, W. Li, F. Meng, L. Wang, H. Dong, and Y. Qiu, *J. Mater. Chem. A* 2, 705–710 (2014).

[43] J. Burschka, N. Pellet, S.-J. Moon, R. Humphry-Baker, P. Gao, M.K. Nazeeruddin, and M. Grätzel, *Nature* 499, 316–319 (2013).

[44] S.N. Habisreutinger, T. Leijtens, G.E. Eperon, S.D. Stranks, R.J. Nicholas, and H.J. Snaith, *Nano Lett.* 14, 5561–5568 (2014).

[45] N. Ahn, K. Kwak, M.S. Jang, H. Yoon, B.Y. Lee, J.-K. Lee, P.V. Pikhitsa, J. Byun, and M. Choi, *Nat. Commun.* 7, 13422 (2016).

[46] J.F. Galisteo-López, M. Anaya, M.E. Calvo, and H. Míguez, *J. Phys. Chemi. Lett.* 6, 2200–2205 (2015).

[47] A. Merdasa, M. Bag, Y. Tian, E. Källman, A. Dobrovolsky, and I.G. Scheblykin, *J. Phys. Chem. C* 120, 10711–10719 (2016).

[48] B.T. Diroll, G. Nedelcu, M.V. Kovalenko, and R.D. Schaller, *Adv. Funct. Mater.* 27, 1606750 (2017).

[49] Y.-H. Kim, C. Wolf, H. Kim, and T.-W. Lee, *Nano Energy* 52, 329–335 (2018).

[50] Y.Y. Li, C.K. Lin, G.L. Zheng, Z.Y. Cheng, H. You, W.D. Wang, and J. Lin, *Chem. Mater.* 18, 3463–3469 (2006).

[51] M.D. Smith, B.A. Connor, and H.I. Karunadasa, *Chem. Rev.* 119, 3104–3139 (2019).

[52] G. Zhou, Z. Liu, J. Huang, M.S. Molokeev, Z. Xiao, C. Ma, and Z. Xia, *J. Phys. Chemi. Lett.* 11, 5956–5962 (2020).

[53] C. Sun, Y.-H. Guo, S.-S. Han, J.-Z. Li, K. Jiang, L.-F. Dong, Q.-L. Liu, C.-Y. Yue, and X.-W. Lei, *Angew. Chem. Int. Ed. Engl.* 59, 16465–16469 (2020).

[54] G. Zhou, X. Jiang, M. Molokeev, Z. Lin, J. Zhao, J. Wang, and Z. Xia, *Chem. Mater.* 31, 5788–5795 (2019).

[55] B.-B. Cui, Y. Han, B. Huang, Y. Zhao, X. Wu, L. Liu, G. Cao, Q. Du, N. Liu, W. Zou, M. Sun, L. Wang, X. Liu, J. Wang, H. Zhou, and Q. Chen, *Nat. Commun.* 10, 5190 (2019).

[56] Y. Jiang, X. Wang, and A. Pan, *Adv. Mater.* 31, 1806671 (2019).

[57] V. D'Innocenzo, G. Grancini, M.J.P. Alcocer, A.R.S. Kandada, S.D. Stranks, M.M. Lee, G. Lanzani, H.J. Snaith, and A. Petrozza, *Nat. Commun.* 5, 3586 (2014).

[58] W. Du, S. Zhang, Z. Wu, Q. Shang, Y. Mi, J. Chen, C. Qin, X. Qiu, Q. Zhang, and X. Liu, *Nanoscale* 11, 3145–3153 (2019).

[59] J.S. Manser, J.A. Christians, and P.V. Kamat, *Chem. Rev.* 116, 12956–13008 (2016).

[60] A.P. Schlaus, M.S. Spencer, K. Miyata, F. Liu, X. Wang, I. Datta, M. Lipson, A. Pan, and X.Y. Zhu, *Nat. Commun.* 10, 265 (2019).

[61] S.A. Bretschneider, I. Ivanov, H.I. Wang, K. Miyata, X. Zhu, and M. Bonn, *Adv. Mater.* 30, 1707312 (2018).

[62] Z. Qin, C. Zhang, L. Chen, T. Yu, X. Wang, and M. Xiao, *Nano Lett.* 21, 7831–7838 (2021).

[63] R. Fan, L. Song, Z. Tang, K. Jia, L. Huang, and W. Bi, *Mater. Lett.* 272, 127878 (2020).

[64] G. Xing, N. Mathews, S.S. Lim, N. Yantara, X. Liu, D. Sabba, M. Grätzel, S. Mhaisalkar, and T.C. Sum, *Nat. Mater.* 13, 476–480 (2014).

[65] H. Zhu, Y. Fu, F. Meng, X. Wu, Z. Gong, Q. Ding, M.V. Gustafsson, M.T. Trinh, S. Jin, and X.Y. Zhu, *Nat. Mater.* 14, 636–642 (2015).

[66] S. Chen, C. Zhang, J. Lee, J. Han, and A. Nurmikko, *Adv. Mater.* 29, 1604781 (2017).

[67] Y. Fang, A. Armin, P. Meredith, and J. Huang, *Nat. Photo.* 13, 1–4 (2019).

[68] X. Gong, M. Tong, Y. Xia, W. Cai, S. Moon Ji, Y. Cao, G. Yu, C.-L. Shieh, B. Nilsson, and J. Heeger Alan, *Science* 325, 1665–1667 (2009).

[69] M.I. Saidaminov, A.L. Abdelhady, B. Murali, E. Alarousu, V.M. Burlakov, W. Peng, I. Dursun, L. Wang, Y. He, G. Maculan, A. Goriely, T. Wu, O.F. Mohammed, and O.M. Bakr, *Nat. Commun.* 6, 7586 (2015).

[70] M.I. Saidaminov, V. Adinolfi, R. Comin, A.L. Abdelhady, W. Peng, I. Dursun, M. Yuan, S. Hoogland, E.H. Sargent, and O.M. Bakr, *Nat. Commun.* 6, 8724 (2015).

[71] J. Song, L. Xu, J. Li, J. Xue, Y. Dong, X. Li, and H. Zeng, *Adv. Mater.* 28, 4861–4869 (2016).

[72] W. Deng, X. Zhang, L. Huang, X. Xu, L. Wang, J. Wang, Q. Shang, S.-T. Lee, and J. Jie, *Adv. Mater.* 28, 2201–2208 (2016).

[73] H. Wu, Y. Ge, G. Niu, and J. Tang, *Matter* 4, 144–163 (2021).

[74] S. Wu, C. Liang, J. Zhang, Z. Wu, X.-L. Wang, R. Zhou, Y. Wang, S. Wang, D.-S. Li, and T. Wu, *Angew. Chem. Int. Ed. Engl.* 59, 18605–18610 (2020).

[75] X. Zheng, W. Zhao, P. Wang, H. Tan, M.I. Saidaminov, S. Tie, L. Chen, Y. Peng, J. Long, and W.-H. Zhang, *J. Energy. Chem.* 49, 299–306 (2020).

[76] L. Li, H. Chen, Z. Fang, X. Meng, C. Zuo, M. Lv, Y. Tian, Y. Fang, Z. Xiao, C. Shan, Z. Xiao, Z. Jin, G. Shen, L. Shen, and L. Ding, *Adv. Mater.* 32, 1907257 (2020).

[77] R.T. Williams, W.W. Wolszczak, X. Yan, and D.L. Carroll, *ACS Nano.* 14, 5161–5169 (2020).

[78] D.M. Panneerselvam and M.Z. Kabir, *J. Mater. Sci-Mater. El.* 28, 7083–7090 (2017).

[79] J.H. Heo, D.H. Shin, J.K. Park, D.H. Kim, S.J. Lee, and S.H. Im, *Adv. Mater.* 30, 1801743 (2018).

[80] Y. Huang, L. Qiao, Y. Jiang, T. He, R. Long, F. Yang, L. Wang, X. Lei, M. Yuan, and J. Chen, *Angew. Chem. Int. Ed. Engl.* 58, 17834–17842 (2019).

[81] Y.C. Kim, K.H. Kim, D.-Y. Son, D.-N. Jeong, J.-Y. Seo, Y.S. Choi, I.T. Han, S.Y. Lee, and N.-G. Park, *Nature* 550, 87–91 (2017).

[82] P. Lecoq, *Nucl. Instrum. Methods Phys. Res. A: Accel. Spectrom. Detect. Assoc. Equip.* 809, 130–139 (2016).

[83] Q. Chen, J. Wu, X. Ou, B. Huang, J. Almutlaq, A.A. Zhumekenov, X. Guan, S. Han, L. Liang, Z. Yi, J. Li, X. Xie, Y. Wang, Y. Li, D. Fan, D.B.L. Teh, A.H. All, O.F. Mohammed, O.M. Bakr, T. Wu, M. Bettinelli, H. Yang, W. Huang, and X. Liu, *Nature* 561, 88–93 (2018).

[84] W. Ma, T. Jiang, Z. Yang, H. Zhang, Y. Su, Z. Chen, X. Chen, Y. Ma, W. Zhu, X. Yu, H. Zhu, J. Qiu, X. Liu, X. Xu, and Y. Yang, *Adv. Sci.* 8, 2003728 (2021).

[85] K. Han, K. Sakhatskyi, J. Jin, Q. Zhang, M.V. Kovalenko, and Z. Xia, *Adv. Mater.* 34, 2110420 (2022).

[86] L.-J. Xu, X. Lin, Q. He, M. Worku, and B. Ma, *Nat. Commun.* 11, 4329 (2020).

[87] M. Zhang, J. Zhu, B. Yang, G. Niu, H. Wu, X. Zhao, L. Yin, T. Jin, X. Liang, and J. Tang, *Nano Lett.* 21, 1392–1399 (2021).

[88] Z. Li, F. Zhou, H. Yao, Z. Ci, Z. Yang, and Z. Jin, *Mater. Today* 48, 155–175 (2021).

[89] Q. Wang, X. Zhang, Z. Jin, J. Zhang, Z. Gao, Y. Li, and S.F. Liu, *ACS Energy Lett.* 2, 1479–1486 (2017).

[90] Y.-C. Zhao, W.-K. Zhou, X. Zhou, K.-H. Liu, D.-P. Yu, and Q. Zhao, *Light Sci. Appl.* 6, e16243–e16243 (2017).

[91] R. Zhuang, X. Wang, W. Ma, Y. Wu, X. Chen, L. Tang, H. Zhu, J. Liu, L. Wu, W. Zhou, X. Liu, and Y. Yang, *Nat. Photo.* 13, 602–608 (2019).

[92] Q. Wang, Z. Jin, D. Chen, D. Bai, H. Bian, J. Sun, G. Zhu, G. Wang, and S. Liu, *Adv. Energy Mater.* 8, 1800007 (2018).

[93] G. Zhou, B. Su, J. Huang, Q. Zhang, and Z. Xia, *Mater. Sci. Eng. R Rep.* 141, 100548 (2020).

[94] M. Li, J. Zhou, M.S. Molokeev, X. Jiang, Z. Lin, J. Zhao, and Z. Xia, *Inorg. Chem.* 58, 13464–13470 (2019).

[95] L. Zhou, J.F. Liao, Z.G. Huang, J.H. Wei, X.D. Wang, W.G. Li, H.Y. Chen, D.B. Kuang, and C.Y. Su, *Angew. Chem. Int. Ed. Engl.* 58, 5277–5281 (2019).

[96] S. Yakunin, B.M. Benin, Y. Shynkarenko, O. Nazarenko, M.I. Bodnarchuk, D.N. Dirin, C. Hofer, S. Cattaneo, and M.V. Kovalenko, *Nat. Mater.* 18, 846–852 (2019).

[97] Q. Wang and K. Domen, *Chem. Rev.* 120, 919–985 (2020).

[98] H. Huang, B. Pradhan, J. Hofkens, M.B.J. Roeffaers, and J.A. Steele, *ACS Energy Lett.* 5, 1107–1123 (2020).

[99] J. Hou, S. Cao, Y. Wu, Z. Gao, F. Liang, Y. Sun, Z. Lin, and L. Sun, *Chem. Eur. J.* 23, 9481–9485 (2017).

[100] Y.-F. Xu, M.-Z. Yang, B.-X. Chen, X.-D. Wang, H.-Y. Chen, D.-B. Kuang, and C.-Y. Su, *J. Am. Chem. Soc.* 139, 5660–5663 (2017).

[101] Z. Zhang, Y. Liang, H. Huang, X. Liu, Q. Li, L. Chen, and D. Xu, *Angew. Chem. Int. Ed. Engl.* 58, 7263–7267 (2019).

[102] X. Zhu, Y. Lin, Y. Sun, M.C. Beard, and Y. Yan, *J. Am. Chem. Soc.* 141, 733–738 (2019).

[103] H. Wang, X. Wang, R. Chen, H. Zhang, X. Wang, J. Wang, J. Zhang, L. Mu, K. Wu, F. Fan, X. Zong, and C. Li, *ACS Energy Lett.* 4, 40–47 (2019).

Index

www.ingramcontent.com/pod-product-compliance
Lightning Source LLC
Chambersburg PA
CBHW060249220326
41598CB00027B/4034